SIMPLIFIED HIGHER ALGEBRA AND DIFFERENTIAL CALCULUS

The Ultimate Guide to Mastering Higher Algebra and Differential Calculus

Kingsley Augustine

Printed by Amazon KDP

Table of Contents

PREFACE

Simplified Higher Algebra and Differential Calculus is a book that contains topics under algebra and differential calculus. This book serves as a useful companion for students in high schools, colleges and universities. It is a valuable textbook for students who want to write entrance test or examination into colleges and universities. This book consists of step-by-step explanation of topics presented in a way that is easy for students to understand. It contains very many worked examples and many self-assessment exercises to ensure that students get a mastery of each topic covered. The answers to the exercises are provided at the end of the book.

What makes this book a unique mathematical asset, is its detailed step by step approach in explaining the topics covered in these branch of mathematics. Instead of solving questions by going straight to the point, leaving you confused and frustrated, this textbook teaches you in simple English, explaining each step taken at a time. Thus, allowing anyone, regardless of their experience in algebra and differential calculus, to understand each topic with ease, and hence make mathematics more interesting.

I give all thanks and Glory to God Almighty, for giving me the grace to write this book. I also wish to express my deep appreciation to my wife Mrs. Mercy Augustine for her patience, understanding and encouragement when I was writing this book. I also thank my children, Dora, Merit and Elvis for their moral support.

Kingsley Augustine.

CHAPTER 1
SIMPLIFICATION OF ALGEBRAIC FRACTIONS

When simplifying single algebraic fractions, factorize the numerator and denominator and then cancel out common terms. However, when two or more algebraic fractions are to be simplified, we treat them like the usual way of dealing with fractions by taking LCM and following the necessary steps.

Examples

1. Simplify $\dfrac{a^2 + ab}{a^2 + ac}$

Solution

Factorize the fraction to give:

$$\frac{a^2 + ab}{a^2 + ac} = \frac{a(a + b)}{a(a + c)}$$

The 'a' will cancel out to give the final answer as follows:

$$\frac{\cancel{a}(a + b)}{\cancel{a}(a + c)} = \frac{a + b}{a + c}$$

Note that we can only cancel out terms linking one another by multiplication. Terms that are linked together by addition sign or subtraction sign cannot be cancelled out. For example, in example 1 above, the fraction $\dfrac{a(a + b)}{a(a + c)}$ can also be written as: $\dfrac{a \times (a + b)}{a \times (a + c)}$. The 'a's linked by multiplication sign can

be cancelled out as shown in our solution above. However, in the final answer given by $\dfrac{a + b}{a + c}$, the a's cannot be cancelled out to give $\dfrac{b}{c}$ since they are linked by addition sign.

2. Simplify $\dfrac{c^2 - 2c - 15}{c^2 - 3c - 10}$

Solution

The numerator and denominator are quadratic expressions. We factorize them as follows:

$$\frac{c^2 - 2c - 15}{c^2 - 3c - 10} = \frac{(c - 5)(c + 3)}{(c - 5)(c + 2)}$$

Therefore, $(c - 5)$ will cancel out to give our final answer as follows:

$$\frac{\cancel{(c - 5)}(c + 3)}{\cancel{(c - 5)}(c + 2)} = \frac{(c + 3)}{(c + 2)}$$

Refer to my book 'Simplified Basic Algebra' for explanations on factorization of quadratic expression in order to understand how the factorization above was carried out.

3. Simplify $\dfrac{8 - 2a - a^2}{2a^2 - 3a - 2}$

<u>Solution</u>

$$\dfrac{8 - 2a - a^2}{2a^2 - 3a - 2}$$

Let us factorize $8 - 2a - a^2$ as follows:

Multiply the first and last terms to give:

$8(-a^2) = -8a^2$.

Two numbers in 'a' whose product is $-8a^2$ and sum is $-2a$ are $-4a$ and $2a$. Replace $-2a$ in the original expression with these two terms. This gives:

$8 - 4a + 2a - a^2$

We now factorize by grouping as follows:

$8 - 4a + 2a - a^2 = 4(2 - a) + a(2 - a)$

$\qquad\qquad\qquad = (2 - a)(4 + a)$ (After using $(2 - a)$ as a common term of the expression)

Similarly, we factorize the denominator as follows:

$2a^2 - 3a - 2 = 2a^2 - 4a + a - 2$

$\qquad\qquad\quad = 2a(a - 2) + 1(a - 2)$

$\qquad\qquad\quad = (a - 2)(2a + 1)$

$\therefore\qquad \dfrac{8 - 2a - a^2}{2a^2 - 3a - 2} = \dfrac{(2-a)(4+a)}{(a - 2)(2a + 1)}$

$\qquad\qquad\qquad = \dfrac{-(a - 2)(4+a)}{(a - 2)(2a + 1)}$ [Note that $(2 - a)$ can also be expressed as $-(a - 2)$]

Therefore, $(a - 2)$ will cancel out to give:

$\dfrac{-(a - 2)(4+a)}{(a - 2)(2a + 1)} = \dfrac{-\cancel{(a - 2)}(4 + a)}{\cancel{(a - 2)}(2a + 1)}$

$\qquad\qquad\qquad = \dfrac{-(4 + a)}{(2a + 1)}$

Note that an expression such as $(a - 2)$ can be converted to $-(2 - a)$. This is done by simply putting a negative sign outside the bracket and changing the signs of the two terms inside the bracket. Similarly, $(5 - x)$ can also be changed to $-(x - 5)$. We only carry out this kind of conversion in order to make it look the same like another term, so that the two terms can be canceled out during the simplification.

4. Simplify: $\dfrac{9a^2 - m^2}{m^2 - 2am - 3a^2}$

<u>Solution</u>

The numerator is a difference of two squares. Recall that a difference of two squares is factorized as follows:

$$a^2 - b^2 = (a + b)(a - b)$$

Similarly, $9a^2 - m^2 = (3a)^2 - m^2$

$$= (3a + m)(3a - m)$$

In order to factorize the denominator, we multiply the first and last terms, since it is a quadratic expression. This gives:

$$m^2(-3a^2) = -3a^2m^2$$

Two numbers in 'am' whose product is $-3a^2m^2$ and sum is $-2am$ are $-3am$ and am.

$$\therefore \quad m^2 - 2am - 3a^2 = m^2 - 3am + am - 3a^2$$

$$= m(m - 3a) + a(m - 3a)$$

$$= (m - 3a)(m + a)$$

$$\therefore \quad \frac{9a^2 - m^2}{m^2 - 2am - 3a^2} = \frac{(3a + m)(3a - m)}{(m - 3a)(m + a)}$$

$$= \frac{-(3a + m)(m - 3a)}{(m - 3a)(m + a)} \qquad \text{(Note that } (3a - m) = -(m - 3a) \text{, as shown)}$$

Hence, $(m - 3a)$ cancels out as follows:

$$\frac{-(3a + m)(m - 3a)}{(m - 3a)(m + a)} = \frac{-(3a + m)\cancel{(m - 3a)}}{\cancel{(m - 3a)}(m + a)}$$

$$= \frac{-(3a + m)}{(m + a)}$$

5. Simplify: $\dfrac{a^2 - am - an + mn}{a^2 - am + an - mn}$

<u>Solution</u>

Factorize both numerator and denominator by grouping terms. This gives:

$$\frac{a^2 - am - an + mn}{a^2 - am + an - mn} = \frac{a(a - m) - n(a - m)}{a(a - m) + n(a - m)}$$

$$= \frac{(a - m)(a - n)}{(a - m)(a + n)}$$

Therefore, $(a - m)$ cancels out as follows:

$$\frac{(a - m)(a - n)}{(a - m)(a + n)} = \frac{\cancel{(a - m)}(a - n)}{\cancel{(a - m)}(a + n)}$$

$$= \frac{(a - n)}{(a + n)}$$

6. Simplify $\dfrac{a+2}{a} - \dfrac{1}{3ab}$

Solution

$$\dfrac{a+2}{a} - \dfrac{1}{3ab}$$

The LCM of a and 3ab is 3ab. Hence divide the LCM by each of the denominator and multiply the value obtained by the respective numerator. This means that 3ab ÷ a = 3b. Then multiply 3b by a + 2 (the numerator). Similarly, 3ab ÷ 3ab = 1. Then multiply 1 by 1(the numerator). Finally divide these multiplied terms by the LCM. This is as shown below.

$$\dfrac{a+2}{a} - \dfrac{1}{3ab} = \dfrac{3b(a+2) - 1(1)}{3ab}$$

$$= \dfrac{3ab + 6b - 1}{3ab}$$

7. Simplify: $\dfrac{1}{4x-2y} - \dfrac{1}{y-2x}$

Solution

$$\dfrac{1}{4x-2y} - \dfrac{1}{y-2x}$$

A careful look at one of the denominator (i.e. $4x - 2y$), shows that it can be factorized and made to look like the other denominator. This is carried out as follows.

$$\dfrac{1}{4x-2y} - \dfrac{1}{y-2x} = \dfrac{1}{2(2x-y)} - \dfrac{1}{y-2x}$$

$$= \dfrac{1}{2(2x-y)} - \dfrac{1}{-(2x-y)} \qquad \text{[Note that } (y-2x) \text{ is also } -(2x-y)]$$

$$= \dfrac{1}{2(2x-y)} + \dfrac{1}{(2x-y)} \qquad \text{(Note that } -(-) \text{ has become +)}$$

The LCM of $2(2x-y)$ and $(2x-y)$ is $2(2x-y)$. Hence we continue our simplification as follows:

$$\dfrac{1}{2(2x-y)} + \dfrac{1}{(2x-y)} = \dfrac{1+2}{2(2x-y)}$$

$$= \dfrac{3}{2(2x-y)}$$

8. Simplify $\dfrac{3}{2ab} + \dfrac{4}{3bc}$

Solution

$$\frac{3}{2ab} + \frac{4}{3bc}$$

The LCM of 2ab and 3bc is 6abc. Therefore:

$$\frac{3}{2ab} + \frac{4}{3bc} = \frac{3c(3) + 2a(4)}{6abc}$$

$$= \frac{9c + 8a}{6abc}$$

9. Simplify $\dfrac{3x}{x-1} - \dfrac{4}{x+2}$

Solution

$$\frac{3x}{x-1} - \frac{4}{x+2}$$

The LCM of $(x-1)$ and $(x+2)$ is $(x-1)(x+2)$. We now simplify as follows:

$$\frac{3x}{x-1} - \frac{4}{x+2} = \frac{(x+2)3x - 4(x-1)}{(x-1)(x+2)}$$

$$= \frac{3x^2 + 6x - 4x + 4}{(x-1)(x+2)}$$

$$= \frac{3x^2 + 2x + 4}{(x-1)(x+2)}$$

10. Simplify $\dfrac{1}{3xy} + \dfrac{5}{4y^2z} - \dfrac{3y}{2x^3}$

<u>Solution</u>

$$\frac{1}{3xy} + \frac{5}{4y^2z} - \frac{3y}{2x^3}$$

The LCM of $3xy$, $4y^2z$ and $2x^3$ is $12x^3y^2z$. Hence, we use this LCM to simplify the expression as follows:

$$\frac{1}{3xy} + \frac{5}{4y^2z} - \frac{3y}{2x^3} = \frac{1(4x^2yz) + 5(3x^3) - 3y(6y^2z)}{12x^3y^2z}$$

$$= \frac{4x^2yz + 15x^3 - 18y^3z}{12x^3y^2z}$$

11. Simplify $\dfrac{3y}{y^2-z^2} - \dfrac{3z}{z^2-y^2}$

<u>Solution</u>

$$\frac{3y}{y^2-z^2} - \frac{3z}{z^2-y^2}$$

The denominators are difference of two squares. Let us factorize the denominators and make them look alike.

$$\frac{3y}{y^2-z^2} - \frac{3z}{z^2-y^2} = \frac{3y}{(y-z)(y+z)} - \frac{3z}{(z-y)(z+y)}$$

$$= \frac{3y}{(y-z)(y+z)} - \frac{3z}{-(y-z)(y+z)}$$

Note that $(z-y)$ is also $-(y-z)$. Note also that $(z+y)$ is the same as $(y+z)$ as rearranged above. Hence the $-(-)$ will become + as follows:

$$\frac{3y}{(y-z)(y+z)} + \frac{3z}{(y-z)(y+z)}$$

The LCM of $(y-z)(y+z)$ and $(y-z)(y+z)$ is simply $(y-z)(y+z)$. When terms are the same, just take one of them as the LCM. Hence, we now continue with the simplification as follows:

$$\frac{3y}{(y-z)(y+z)} + \frac{3z}{(y-z)(y+z)} = \frac{3y(1)+3z(1)}{(y-z)(y+z)}$$ (Note that $(y-z)(y+z) \div (y-z)(y+z) = 1$)

$$= \frac{3y+3z}{(y-z)(y+z)}$$

$$= \frac{3(y+z)}{(y-z)(y+z)}$$

$$= \frac{3\cancel{(y+z)}}{(y-z)\cancel{(y+z)}}$$

$$= \frac{3}{(y-z)}$$

Exercise 1

1. Simplify $\dfrac{m^2+md}{m^2+mf}$

2. Simplify $\dfrac{x^2-4x-21}{x^2-x-12}$

3. Simplify $\dfrac{6-x-x^2}{3x^2+4x-15}$

4. Simplify $\dfrac{16m^2 - n^2}{n^2 - 3mn - 4m^2}$

5. Simplify: $\dfrac{p^2 - pq - pr + qr}{p^2 - pq + pr - qr}$

6. Simplify $\dfrac{2a + 3}{b} - \dfrac{2}{3b}$

7. Simplify: $\dfrac{3}{10x - 5y} - \dfrac{2}{y - 2x}$

8. Simplify $\dfrac{1}{6mn} + \dfrac{3}{2n}$

9. Simplify $\dfrac{5a}{a - 2} - \dfrac{2}{a + 3}$

10. Simplify $\dfrac{2}{8ab} + \dfrac{1}{3a^2b} - \dfrac{3a}{12b^3}$

11. Simplify $\dfrac{4}{9m^2 - 4n^2} - \dfrac{5n}{4n^2 - 9m^2}$

CHAPTER 2
EQUATIONS AND SUBSTITUTIONS INVOLVING FRACTIONS

Examples

1. Solve the equation $\dfrac{3}{2b-5} - \dfrac{4}{b-3} = 0$

<u>Solution</u>

$$\frac{3}{2b-5} - \frac{4}{b-3} = 0$$

Multiply each term by the LCM, i.e. $(2b-5)(b-3)$. This gives:

$$(2b-5)(b-3)\frac{3}{2b-5} - (2b-5)(b-3)\frac{4}{b-3} = (2b-5)(b-3)0$$

Therefore, we cancel the common denominators as follows:

$\cancel{(2b-5)}(b-3)\dfrac{3}{\cancel{2b-5}} - (2b-5)\cancel{(b-3)}\dfrac{4}{\cancel{b-3}} = 0$ (Note that zero multiplies a number to give zero)

$(b-3)3 - (2b-5)4 = 0$

$3b - 9 - 8b + 20 = 0$ (Take note of the negative sign outside the second bracket)

$3b - 8b = 9 - 20$

$-5b = -11$

$b = \dfrac{-11}{-5}$

$b = 2\dfrac{1}{5}$ (Note that the negative sign has cancelled out)

2. Solve the equation $\dfrac{a-4}{7} = \dfrac{2}{3a-1}$

<u>Solution</u>

$$\frac{a-4}{7} = \frac{2}{3a-1}$$

We can multiply each term in the equation by the LCM, i.e. $7(3a-1)$. However, since there is a single fraction on each side of the equation, a simple way to solve this kind of equation is to cross multiply. This gives:

$(a-4)(3a-1) = 7(2)$

$3a^2 - a - 12a + 4 = 14$

$3a^2 - 13a + 4 - 14 = 0$

$3a^2 - 13a - 10 = 0$

Solving this quadratic equation by factorization gives:

$3a^2 - 15a + 2a - 10 = 0$

Note that −15a and +2a are the two numbers whose product will give $-30a^2$ (i.e. $3a^2 \times (-10) = -30a^2$) and whose sum will give −13a. Hence they are used to substitute for −13a in the original quadratic equation. Therefore, we factorize by grouping as follows:

$$3a(a-5) + 2(a-5) = 0$$
$$(a-5)(3a+2) = 0$$
$$\therefore \quad a = 5 \text{ or } a = -\frac{2}{3}$$

3. Solve the equation $\dfrac{3}{x-4} = \dfrac{2}{x-1} - 4$

Solution

$$\frac{3}{x-4} = \frac{2}{x-1} - 4$$

Multiply each term by the LCM, i.e. $(x-4)(x-1)$. This gives:

$$(x-4)(x-1)\frac{3}{x-4} = (x-4)(x-1)\frac{2}{x-1} - (x-4)(x-1)4$$

We now cancel out the common terms as follows:

$$\cancel{(x-4)}(x-1)\frac{3}{\cancel{x-4}} = (x-4)\cancel{(x-1)}\frac{2}{\cancel{x-1}} - (x-4)(x-1)4$$

$$(x-1)3 = (x-4)2 - 4(x-4)(x-1)$$
$$3x - 3 = 2x - 8 - 4(x^2 - x - 4x + 4)$$
$$3x - 3 = 2x - 8 - 4x^2 + 4x + 16x - 16$$
$$3x - 3 = 22x - 24 - 4x^2$$
$$4x^2 + 3x - 22x - 3 + 24 = 0$$
$$4x^2 - 19x + 21 = 0$$

Solving this quadratic equation by factorization gives:

$$4x^2 - 12x - 7x + 21 = 0$$
$$4x(x-3) - 7(x-3) = 0$$
$$(x-3)(4x-7) = 0$$
$$x = 3 \text{ or } x = \frac{7}{4}$$
$$x = 3 \text{ or } x = 1\frac{3}{4}$$

Undefined Fractions

A fraction is undefined if the denominator of the fraction is zero.

Examples

1. Find the value of x for which the fraction $\dfrac{8}{15 + 3x}$ is undefined.

15

Solution

$$\frac{8}{15 + 3x}$$

For this fraction to be undefined, the denominator must be equal to zero. Hence:

$15 + 3x = 0$ (Note that only the denominator is equated to zero)

$3x = -15$

$x = \dfrac{-15}{3}$

$x = -5$

$\therefore$ The fraction is undefined when $x = -5$

2. Find the value of x for which the fraction $\dfrac{6x}{2x - 5}$ is not defined.

Solution

For this fraction not to be defined, the denominator must be equal to zero. Hence:

$2x - 5 = 0$

$2x = 5$

$x = \dfrac{5}{2}$

$x = 2\dfrac{1}{2}$

$\therefore$ The fraction is not defined when $x = 2\dfrac{1}{2}$

3. For what value(s) of b is the fraction $\dfrac{2b + 11}{b^2 + b - 20}$

a. not defined

b. equal to zero?

Solution

a. $\dfrac{2b + 11}{b^2 + b - 20}$

For this fraction not to be defined, we equate the denominator to zero as follows:

$b^2 + b - 20 = 0$

Solving this quadratic equation by factorization gives:

$(b + 5)(b - 4) = 0$

Therefore, b = −5 or b = 4

Hence, the expression is not defined when b = −5 or b = 4

b. Recall that a fraction is zero when the numerator of the fraction is zero. Therefore we now equate the numerator to zero as follows:

$$2b + 11 = 0$$
$$2b = -11$$
$$b = \frac{-11}{2}$$
$$b = -5\frac{1}{2}$$

The fraction is equal to zero when $b = -5\frac{1}{2}$

4. For what value(s) of x is the expression $\dfrac{x^2 + 3x + 2}{2x + 3}$

a. undefined

b. zero?

Solution

a. $\dfrac{x^2 + 3x + 2}{2x + 3}$

For this expression to be undefined:
$$2x + 3 = 0$$
$$2x = -3$$
$$x = \frac{-3}{2}$$

Therefore, $x = -1\frac{1}{2}$

Hence, the expression is undefined when $x = -1\frac{1}{2}$

b. This expression is zero if:
$$x^2 + 3x + 2 = 0$$

Solving this quadratic equation by factorization gives:
$$x^2 + 3x + 2 = 0$$
$$(x + 1)(x + 2) = 0$$
$$x = -1 \text{ or } x = -2$$

∴ The expression is zero when $x = -1$ or $x = -2$

5. For what values of x is the expression $\dfrac{2x^2 + 3x - 2}{3x^2 - 14x + 8}$

a. zero

b. undefined?

<u>Solution</u>

a. $\dfrac{2x^2 + 3x - 2}{3x^2 - 14x + 8}$

This expression is zero if the numerator is zero. Therefore, we equate it to zero as follows:

$$2x^2 + 3x - 2 = 0$$

Let us solve this quadratic equation by factorization as follows:

$$2x^2 + 3x - 2 = 0$$
$$2x^2 + 4x - x - 2 = 0$$
$$2x(x + 2) - 1(x + 2) = 0$$
$$(x + 2)(2x - 1) = 0$$
$$x = -2 \text{ or } x = \dfrac{1}{2} \qquad \text{(After equating each bracket to zero and solving for } x\text{)}$$

∴ The expression is zero when $x = -2$ or $x = \dfrac{1}{2}$

b. The expression is undefined if the denominator is zero. Therefore, we equate it to zero as follows:

$$3x^2 - 14x + 8 = 0$$

Let us solve this quadratic equation by factorization as follows:

$$3x^2 - 14x + 8 = 0$$
$$3x^2 - 12x - 2x + 8 = 0$$
$$3x(x - 4) - 2(x - 4) = 0$$
$$(x - 4)(3x - 2) = 0$$
$$x = 4 \text{ or } x = \dfrac{2}{3} \qquad \text{(After equating each bracket to zero and solving for } x\text{)}$$

∴ The expression is undefined when $x = 4$ or $x = \dfrac{2}{3}$

Substitution in Algebraic Fractions

Examples

1. If $y = \dfrac{a^2 - b}{c - b}$, find the value of y when a = −2, b = −1 and c = 4

<u>Solution</u>

$$y = \dfrac{a^2 - b}{c - b}$$

We simply substitute the various given values of a, b and c into the equation in order to obtain the value of y. This gives:

$$y = \dfrac{a^2 - b}{c - b}$$

$$= \frac{(-2)^2 - (-1)}{4 - (-1)}$$

$$= \frac{4 + 1}{4 + 1}$$

$$= \frac{5}{5}$$

$$y = 1$$

2. If $x = \frac{m^3 + 5n - p}{2mp + m - 3n^2}$, find the value of x when m = –1, n = 3 and p = 6

Solution

$$x = \frac{m^3 + 5n - p}{2mp + m - 3n^2}$$

We simply substitute the various given values of m, n and p into the equation in order to obtain the value of x. This gives:

$$x = \frac{m^3 + 5n - p}{2mp + m - 3n^2}$$

$$= \frac{(-1)^3 + 5(3) - 6}{2(-1)(6) + (-1) - 3(3)^2}$$

$$= \frac{-1 + 15 - 6}{-12 - 1 - 27}$$

$$= \frac{8}{-40}$$

$$x = -\frac{1}{5}$$

3. If p : q = 9 : 5, evaluate $\frac{15p - 2q}{5p + 16q}$

Solution

An easy way of solving this problem is to substitute 9 for p and 5 for q in the given expression.

$$\therefore \frac{15p - 2q}{5p + 16q} = \frac{15(9) - 2(5)}{5(9) + 16(5)}$$

$$= \frac{135 - 10}{45 + 80}$$

$$= \frac{125}{125}$$

$$= 1$$

4. If $a : b = 5 : 3$, evaluate $\dfrac{6a + b}{a - \frac{1}{3}b}$

Solution

An easy way of solving this problem is to substitute 5 for a and 3 for b in the given expression.

$$\therefore \quad \frac{6a + b}{a - \frac{1}{3}b} = \frac{6(5) + 3}{5 - \frac{1}{3}(3)}$$

$$= \frac{30 + 3}{5 - 1}$$

$$= \frac{33}{4}$$

$$= 8\frac{1}{4}$$

5. If $\dfrac{m}{n} = \dfrac{3}{4}$, evaluate $\dfrac{7m + n}{2m - \frac{1}{5}n}$

Solution

An easy way of solving this problem is to substitute 3 for m and 4 for q in the given expression.

$$\therefore \quad \frac{7m + n}{2m - \frac{1}{5}n} = \frac{7(3) + 4}{2(3) - \frac{1}{5}(4)}$$

$$= \frac{21 + 4}{6 - \frac{4}{5}}$$

$$= \frac{25}{\frac{30 - 4}{5}}$$

$$= \frac{25}{\frac{26}{5}}$$

$$= \frac{25}{1} \times \frac{5}{26}$$

$$= \frac{125}{26}$$

$$= 4\frac{21}{26}$$

6. If $a = \frac{x+2}{x-1}$, express $\frac{2a+3}{a-1}$ in terms of x

Solution

In order to express $\frac{2a+3}{a-1}$ in terms of x, we simply substitute the expression for 'a' in $\frac{2a+3}{a-1}$.

Since 'a' is expressed in terms x, then $\frac{2a+3}{a-1}$ will become expressed in x if 'a' is substituted in it.

Therefore, let us substitute $\frac{x+2}{x-1}$ for 'a' in $\frac{2a+3}{a-1}$. This gives:

$$\frac{2a+3}{a-1} = \frac{2\left(\frac{x+2}{x-1}\right)+3}{\frac{x+2}{x-1}-1}$$

$$= \frac{\left(\frac{2x+4}{x-1}\right)+3}{\frac{x+2}{x-1}-1}$$

$$= \frac{\left(\frac{2x+4+3(x-1)}{x-1}\right)}{\frac{x+2-1(x-1)}{x-1}}$$

$$= \frac{\left(\frac{2x+4+3x-3}{x-1}\right)}{\frac{x+2-x+1)}{x-1}}$$

$$= \frac{\left(\frac{5x+1}{x-1}\right)}{\frac{3}{x-1}}$$

$$= \frac{5x+1}{x-1} \times \frac{x-1}{3}$$

$$= \frac{5x+1}{\cancel{x-1}} \times \frac{\cancel{x-1}}{3}$$

$$= \frac{5x+1}{3}$$

7. If $m = \dfrac{y + 5}{y - 3}$, express $\dfrac{5m + 2}{m - 3}$ in terms of y

Solution

In order to express $\dfrac{5m + 2}{m - 3}$ in terms of y, we simply substitute the expression for m in $\dfrac{5m + 2}{m - 3}$.

Since m is expressed in terms y, then $\dfrac{5m + 2}{m - 3}$ will become expressed in y if m is substituted in it.

Therefore, let us substitute $\dfrac{y + 5}{y - 3}$ for m in $\dfrac{5m + 2}{m - 3}$. This gives:

$$\frac{5m+2}{m-3} = \frac{5\left(\frac{y+5}{y-3}\right) + 2}{\left(\frac{y+5}{y-3}\right) - 3}$$

$$= \frac{\left(\frac{5y+25}{y-3}\right) + 2}{\left(\frac{y+5}{y-3}\right) - 3}$$

$$= \frac{\left(\frac{5y + 25 + 2(y-3)}{y-3}\right)}{\frac{y + 5 - 3(y-3)}{y-3}}$$

$$= \frac{\left(\frac{5y + 25 + 2y - 6}{y-3}\right)}{\frac{y + 5 - 3y + 9)}{y-3}}$$

$$= \frac{\left(\frac{7y + 19}{y-3}\right)}{\frac{14 - 2y}{y-3}}$$

$$= \frac{7y + 19}{y-3} \times \frac{y-3}{14 - 2y}$$

$$= \frac{7y + 19}{\cancel{y-3}} \times \frac{\cancel{y-3}}{14 - 2y}$$

$$= \frac{7y + 19}{14 - 2y}$$

Exercise 2

1. Solve the equation $\dfrac{2}{m-2} - \dfrac{3}{m-5} = 0$

2. Solve the equation $\dfrac{b-1}{5} = \dfrac{5}{2b-7}$

3. Solve the equation $\dfrac{1}{x-5} = \dfrac{9}{2x-12} - 4$

4. Find the value of x for which the fraction $\dfrac{1}{8+x}$ is undefined.

5. Find the value of x for which the fraction $\dfrac{2x}{3x-7}$ is not defined.

6. For what value(s) of a is the fraction $\dfrac{5a+12}{a^2-a-30}$

   a. zero
   b. undefined?

7. For what value(s) of m is the expression $\dfrac{2m^2+8m+8}{5m+9}$

   a. zero
   b. undefined?

8. For what values of x is the expression $\dfrac{5x^2+13x-6}{2x^2-x-15}$

   a. zero
   b. undefined?

9. If m = $\dfrac{p^2-q}{q-r}$, find the value of m when p = −5, q = 3 and r = -2

10. If $x = \dfrac{2a^3+5b-8c}{7ab+b-3c^2}$, find the value of x when a = −2, b = 5 and c = - 4

11. If m : n = 2 : 7, evaluate $\dfrac{6m+4n}{12m-2n}$

12. If a : b = 9 : 5, evaluate $\dfrac{2a+3b}{3a-\frac{1}{5}b}$

CHAPTER 3
SIMULTANEOUS EQUATIONS INVOLVING FRACTIONS

Examples

1. Solve the simultaneous equation: $\dfrac{x}{2} - \dfrac{y}{3} = \dfrac{1}{6}$

$$\dfrac{x}{2} - \dfrac{y}{6} = 5$$

<u>Solution</u>

$$\dfrac{x}{2} - \dfrac{y}{3} = \dfrac{1}{6} \ \text{...................Equation (1)}$$

$$\dfrac{x}{2} - \dfrac{y}{6} = 5 \ \text{...................Equation (2)}$$

This is a case where the variable is in the numerator. In cases like this, we clear out the fractions by multiplying each term by the LCM of the denominators. The LCM in equation (1) and (2) is 6. Hence we clear out the fractions by multiplying each term in equation (1) and (2) by 6. This gives:

$$6(\tfrac{x}{2}) - 6(\tfrac{y}{3}) = 6(\tfrac{1}{6}) \ \text{...................Equation (1)}$$
$$6(\tfrac{x}{2}) - 6(\tfrac{y}{6}) = 6(5) \ \text{...................Equation (2)}$$

These now simplify to give equation 3 and equation 4 respectively, and we solve them simultaneously as shown below:

$$3x - 2y = 1 \ \text{...................Equation (3)}$$
$$\underline{3x - y = 30} \ \text{...................Equation (4)}$$

Equation (4) – equation (3): $y = 29$ (Note that $-y - (-2y) = -y + 2y = y$. Also, $30 - 1 = 29$)

In equation (3), substitute 29 for y in order to obtain x. This gives:

$$3x - 2y = 1 \ \text{...................Equation (3)}$$
$$3x - 2(29) = 1$$
$$3x - 58 = 1$$
$$3x = 1 + 58$$
$$3x = 59$$
$$x = \dfrac{59}{3}$$

Hence, $x = \dfrac{59}{3}$ and y = 29

2. Solve simultaneously: $\dfrac{1}{x} + \dfrac{1}{y} = 5$

$$\dfrac{1}{y} - \dfrac{1}{x} = 1$$

Solution

METHOD 1

$$\dfrac{1}{x} + \dfrac{1}{y} = 5$$

$$\dfrac{1}{y} - \dfrac{1}{x} = 1$$

In these equations the variables are in the denominators. So, let us take $\dfrac{1}{x}$ = a and $\dfrac{1}{y}$ = b. Hence

we can rewrite the equations as follows:

a + b = 5

b − a = 1

Or,

$$a + b = 5 \ \text{....................Equation (1)}$$

$$-a + b = 1 \ \text{..................Equation (2)}$$

Equation (1) + equation (2):   2b = 6

$$b = \dfrac{6}{2}$$

$$b = 3$$

Substitute 3 for b in equation (1). This gives:

a + b = 5 ....Equation (1)

a + 3 = 5

a = 5 − 3

a = 2

Hence a = 2 and b = 3

But recall that $\dfrac{1}{x}$ = a and $\dfrac{1}{y}$ = b, and we are actually solving for x and y. Therefore, we now

substitute the values of a and b appropriately, in order to obtain x and y as follows:

$$\dfrac{1}{x} = a$$

$$\dfrac{1}{x} = 2 \quad \text{(Since a = 2)}$$

When we cross multiply it gives:

2x = 1

$$x = \dfrac{1}{2}$$

Similarly, $\dfrac{1}{y}$ = b

$$\frac{1}{y} = 3 \qquad \text{(Since b = 3)}$$

When we cross multiply it gives:

$$3y = 1$$

$$y = \frac{1}{3}$$

Therefore, $x = \frac{1}{2}$ and $y = \frac{1}{3}$

METHOD 2

$$\frac{1}{x} + \frac{1}{y} = 5$$

$$\frac{1}{y} - \frac{1}{x} = 1$$

Rearranging the equations gives:

$$\frac{1}{x} + \frac{1}{y} = 5 \text{ ....Equation (1)}$$

$$-\frac{1}{x} + \frac{1}{y} = 1 \text{ ....Equation (2)}$$

Since the coefficients of $\frac{1}{x}$ are the same in the two equations, but with different signs, then $\frac{1}{x}$ can be eliminated by using elimination method as follows:

$$\frac{1}{x} + \frac{1}{y} = 5 \text{ ....Equation (1)}$$

$$-\frac{1}{x} + \frac{1}{y} = 1 \text{ ....Equation (2)}$$

Equation (1) + equation (2): $\quad \dfrac{2}{y} = 6 \qquad$ (Note that $\frac{1}{y} + \frac{1}{y} = \frac{2}{y}$ and 5 + 1 = 6)

Hence, $y = \dfrac{2}{6}$

$$y = \frac{1}{3}$$

Substitute $\frac{1}{3}$ for y in equation 2 in order to obtain x. This gives:

$$-\frac{1}{x} + \frac{1}{y} = 1 \text{ ....Equation (2)}$$

$$-\frac{1}{x} + \frac{1}{\frac{1}{3}} = 1$$

$$3 - 1 = \frac{1}{x}$$

$$2 = \frac{1}{x}$$

$$x = \frac{1}{2}$$

Hence, $x = \frac{1}{2}$ and $y = \frac{1}{3}$

3. Solve the simultaneous equations: $\frac{3}{c} - \frac{4}{d} = \frac{1}{3}$

$\frac{2}{c} - \frac{5}{d} = 1$

<u>Solution</u>

$$\frac{3}{c} - \frac{4}{d} = \frac{1}{3} \quad ....Equation\ (1)$$

$$\frac{2}{c} - \frac{5}{d} = 1 \quad ....Equation\ (2)$$

Let us make the coefficient of $\frac{1}{c}$ to be equal in the two equations so that it can be eliminated.

Therefore, multiply equation (1) by 2 and equation (2) by 3. This gives:

$$\frac{6}{c} - \frac{8}{d} = \frac{2}{3} \quad ....Equation\ (3)$$

$$\frac{6}{c} - \frac{15}{d} = 3 \quad ....Equation\ (4)$$

Equation (3) – Equation (4): $\frac{7}{d} = -\frac{7}{3}$

Hence, $d = \frac{7 \times 3}{-7}$

$$d = -3$$

Substitute –3 for d in equation (2) in order to obtain c. This gives:

$$\frac{2}{c} - \frac{5}{d} = 1 \quad ....Equation\ (2)$$

$$\frac{2}{c} - \frac{5}{-3} = 1$$

$$\frac{2}{c} + \frac{5}{3} = 1$$

$$\frac{2}{c} = 1 - \frac{5}{3}$$

$$\frac{2}{c} = \frac{-2}{3}$$

$$c = \frac{2 \times 3}{-2}$$

$$c = -3$$

Therefore, c = –3 and d = –3

4. Solve the simultaneous equations: $\dfrac{a}{2} - \dfrac{b}{5} = 1$

$$b - \frac{a}{3} = 8$$

Solution

Rearranging the equations gives:

$\dfrac{a}{2} - \dfrac{b}{5} = 1$ ....Equation (1)

$-\dfrac{a}{3} + b = 8$ ....Equation (2)

Let us make the coefficient of a to be equal in the two equations so that it can be eliminated.

Therefore, multiply equation (1) by $\dfrac{1}{3}$ and equation (2) by $\dfrac{1}{2}$. This gives:

$\dfrac{a}{6} - \dfrac{b}{15} = \dfrac{1}{3}$ ....Equation (3)

$-\dfrac{a}{6} + \dfrac{b}{2} = 4$ ....Equation (4)

Equation (3) + Equation (4): $\dfrac{13b}{30} = \dfrac{13}{3}$

Hence, b = $\dfrac{30 \times 13}{13 \times 3}$

b = 10

Substitute 10 for b in equation (1) in order to obtain a. This gives:

$\dfrac{a}{2} - \dfrac{b}{5} = 1$ ....Equation (1)

$\dfrac{a}{2} - \dfrac{10}{5} = 1$

$\dfrac{a}{2} = 1 + 2$

a = 3 x 2

a = 6

Hence, a = 6 and b = 10

5. Solve simultaneously the equations: $\dfrac{3}{m} + \dfrac{5}{n} = 1$

$$\dfrac{9}{m} - \dfrac{5}{n} = 1$$

Solution

$$\dfrac{3}{m} + \dfrac{5}{n} = 1 \text{ ....Equation(1)}$$

$$\dfrac{9}{m} - \dfrac{5}{n} = 1 \text{ ....Equation(2)}$$

Let us make the coefficient of $\dfrac{1}{m}$ to be equal in the two equations by multiplying equation (1) by 3. This gives:

$$\dfrac{9}{m} + \dfrac{15}{n} = 3 \text{ ....Equation (3)}$$

$$\dfrac{9}{m} - \dfrac{5}{n} = 1 \text{ ....Equation (4)}$$

Equation (3) – Equation (4): $\qquad \dfrac{20}{n} = 2$

Hence, $n = \dfrac{20}{2}$

$\qquad n = 10$

Substitute 10 for n in equation (1) in order to obtain m. This gives:

$$\dfrac{3}{m} + \dfrac{5}{n} = 1 \text{ ....Equation(1)}$$

$$\dfrac{3}{m} + \dfrac{5}{10} = 1$$

$$\dfrac{3}{m} = 1 - \dfrac{1}{2}$$

$$\dfrac{3}{m} = \dfrac{1}{2}$$

$\qquad m = 3 \times 2$

$\qquad m = 6$

Hence, m = 6 and n = 10

Exercise 3

1. Solve the simultaneous equation:
$$\frac{x}{3} - \frac{y}{4} = -\frac{1}{6}$$
$$\frac{x}{3} - \frac{y}{8} = \frac{1}{12}$$

2. Solve simultaneously, the equation:
$$\frac{1}{a} + \frac{1}{b} = 3$$
$$\frac{1}{a} - \frac{1}{b} = 5$$

3. Solve the simultaneous equations:
$$\frac{2}{c} + \frac{3}{d} = 2$$
$$\frac{3}{c} - \frac{1}{d} = 1\frac{1}{6}$$

4. Solve the simultaneous equations:
$$\frac{a}{5} - \frac{b}{4} = \frac{1}{10}$$
$$a - \frac{b}{2} = 1$$

5. Solve simultaneously the equations:
$$\frac{4}{p} + \frac{6}{q} = 4$$
$$\frac{1}{p} - \frac{3}{q} = -8$$

6. Solve the simultaneous equations:
$$\frac{2}{c} - \frac{10}{d} = 3$$
$$\frac{5}{c} - \frac{5}{d} = 3\frac{1}{2}$$

7. Solve the simultaneous equations:
$$\frac{x}{2} - \frac{y}{4} = \frac{1}{20}$$
$$\frac{x}{4} + \frac{y}{4} = \frac{7}{40}$$

8. Solve simultaneously the equations:
$$\frac{1}{m} + \frac{1}{n} = 2$$
$$\frac{3}{m} - \frac{6}{n} = 5$$

CHAPTER 4
ABSOLUTE VALUE EQUATION (MODULUS EQUATION)

The absolute value of a number is a positive (or zero) value of the number. Whether the number was originally positive or negative, its absolute value must be positive. For example, $|7| = 7$ and $|-7| = 7$. This shows that for each absolute value of a number/expression, there are two possible numbers/expressions. Hence, when solving absolute value equations, we split the equation into two possible equations.

Examples

1. Solve $|x| = 5$

Solution

$$|x| = 5$$

Remove the absolute value bars and split the equation into two cases. The first case should have a positive right hand side, while the second case should have a negative right hand side. This gives:

$$x = 5 \quad \text{or} \quad x = -5$$

2. Solve the following equations:

a. $|3x - 2| = 4$

b. $|2x + 1| = 9$

c. $|7 - 5x| = 2$

Solutions

a. $|3x - 2| = 4$

Simply remove the bars and take a positive right hand value, and then a negative right hand value to obtain two possible equations. Then solve each of the equations separately. This gives:

$$
\begin{array}{ll}
3x - 2 = 4 & \qquad \text{or} \qquad 3x - 2 = -4 \\
3x = 4 + 2 & \qquad\qquad\qquad 3x = -4 + 2 \\
3x = 6 & \qquad\qquad\qquad 3x = -2 \\
x = \dfrac{6}{3} & \qquad\qquad\qquad x = -\dfrac{2}{3} \\
x = 2 &
\end{array}
$$

Hence, $x = 2$ or $x = -\dfrac{2}{3}$

Take note of how the two equations were solved separately.

b. $|2x + 1| = 9$

Remove the bars and take a positive right hand value to form one equation. Form a second equation by also removing the bars and taking a negative right hand value. Then solve each of the equations separately. This gives:

$$2x + 1 = 9 \qquad \text{or} \qquad 2x + 1 = -9$$
$$2x = 9 - 1 \qquad\qquad 2x = -9 - 1$$
$$2x = 8 \qquad\qquad 2x = -10$$
$$x = \frac{8}{2} \qquad\qquad x = \frac{-10}{2}$$
$$x = 4 \qquad\qquad x = -5$$

Hence, $x = 4$ or $x = -5$

c. $|7 - 5x| = 2$

$$7 - 5x = 2 \qquad \text{or} \qquad 7 - 5x = -2$$
$$7 - 2 = 5x \qquad\qquad 7 + 2 = 5x$$
$$5 = 5x \qquad\qquad 9 = 5x$$
$$x = \frac{5}{5} \qquad\qquad 5x = 9$$
$$x = 1 \qquad\qquad x = \frac{9}{5}$$

Hence, $x = 1$ or $x = 1\frac{4}{5}$

3. Solve the following equation:
a. $|4x - 11| = 0$
b. $|2x - 5| = -1$

Solutions

a. $|4x - 11| = 0$

In this case there is only one equation that we can obtain. This is given by:

$$4x - 11 = 0$$
$$4x = 11$$
$$x = \frac{11}{4}$$

Hence, $x = 2\frac{3}{4}$

b. $|2x - 5| = -1$

Recall that an absolute value must be positive. But this equation is telling us that absolute value is negative. This is impossible. An absolute value cannot be negative (i.e. −1 in this case), hence there is no solution to this equation.

4. Solve the following equations:

a. $|x-5| + 8 = 12$

b. $|3x+7| - 14 = -4$

Solution

a. $|x-5| + 8 = 12$

Take the constant term (outside the bars) to the right hand side of the equation. This gives:

$|x-5| = 12 - 8$

$|x-5| = 4$

Let us now remove the bars and split this equation into two possible equations. This gives:

$x-5 = 4$ or $x-5 = -4$

$x = 4 + 5$ $x = -4 + 5$

$x = 9$ $x = 1$

Hence, $x = 9$ or $x = 1$

b. $|3x+7| - 14 = -4$

Do not think that this equation has no solution since there is a negative value on the right hand side. If we simplify the equation we will get a positive value on the right hand side, with the absolute value bars still in place. Therefore, let us simplify the equation as follows.

Take the constant term to the right hand side of the equation. This gives:

$|3x+7| = -4 + 14$

$|3x+7| = 10$

Now that we have a positive value on the right hand side, let us remove the bars and split this equation into two possible equations. This gives:

$3x+7 = 10$ or $3x+7 = -10$

$3x = 10 - 7$ $3x = -10 - 7$

$3x = 3$ $3x = -17$

$x = \dfrac{3}{3}$ $x = -\dfrac{17}{3}$

$x = 1$ $x = -5\dfrac{2}{3}$

Hence, $x = 1$ or $x = -5\dfrac{2}{3}$

5. Solve the following absolute value equations:

a. $|x-6| = 2x - 3$

b. $|5x+2| = x + 4$

c. $|3x-4| = |2x+3|$

<u>Solution</u>

a. $|x-6| = 2x-3$

Let us split the equation into two equations as follows:

$x-6 = 2x-3$ or $x-6 = -(2x-3)$

$x-2x = -3+6$ $x-6 = -2x+3$

$-x = 3$ $x+2x = 3+6$

$x = -3$ $3x = 9$

$$x = \frac{9}{3}$$

$x = 3$

Hence, $x = -3$ or $x = 3$

Now, in equations such as this, it is necessary for us to test the values obtained to see if they are true solutions. In order to do this, we substitute the values of x into the right side of the equation (the side without the absolute value bars) to see if it will give us a positive value. A negative value means that the value of x obtained is not a solution to the equation. Let us write out the part of the equation without the absolute value bar as follows:

$2x-3$

When $x = -3$, the above expression gives:

$2x-3 = 2(-3) - 3$

$= -6-3 = -9$

Hence $x = -3$ is not a solution since it gives a negative right hand side. This is because an absolute value cannot be negative.

When $x = 3$, the above expression gives:

$2x-3 = 2(3) - 3$

$= 6-3 = 3$

Hence $x = 3$ is a solution since it gives a positive value on the right hand side.

Therefore the overall solution of the equation is $x = 3$

b. $|5x+2| = x+4$

Let us split the equation into two equations as follows:

$5x+2 = x+4$ or $5x+2 = -(x+4)$

$5x-x = 4-2$ $5x+2 = -x-4$

$4x = 2$ $5x+x = -4-2$

$$x = \frac{2}{4}$$ $6x = -6$

$$x = \frac{1}{2}$$ $$x = \frac{-6}{6}$$

                              $x = -1$

Let us now find out if these two values of x are solutions of the equations. In order to do this, we substitute the values of x into the right side of the equation (the side without the absolute value bars). Let us write out the part of the equation without the absolute value bar as follows:

$$x + 4$$

When $x = \dfrac{1}{2}$, the above expression gives:

$$x + 4 = \dfrac{1}{2} + 4$$
$$= 4\dfrac{1}{2}$$

Hence $x = \dfrac{1}{2}$ is a solution since it gives a positive right hand side.

When $x = -1$, the above expression gives:

$$x + 4 = -1 + 4$$
$$= 3$$

Hence $x = -1$ is a solution since it gives a positive value on the right hand side.

Therefore the overall solutions of the equation are $x = \dfrac{1}{2}$ or $x = -1$

c. $|3x - 4| = |2x + 3|$

Let us split the equation into two equations as follows:

$3x - 4 = 2x + 3$	or	$3x - 4 = -(2x + 3)$
$3x - 2x = 3 + 4$		$3x - 4 = -2x - 3$
$x = 7$		$3x + 2x = -3 + 4$
		$5x = 1$
		$x = \dfrac{1}{5}$

Since this question has the absolute value bars on both side of the equation, then our two values of x are correct. There is no need to verify out solution.

Therefore, $x = 7$ or $x = \dfrac{1}{5}$

6. Solve the following absolute value equations:

a. $||5x + 2| - 5| = 12$

b. $||7 - 2x| - 8| = 5$

Solution

a. $||5x + 2| - 5| = 12$

This is a case where one absolute value expression is nested in another. Working with the outermost bars, let us split the overall equation into two equations. This gives:

$|5x + 2| - 5 = 12$	or	$|5x + 2| - 5 = -12$
$|5x + 2| = 12 + 5$		$|5x + 2| = -12 + 5$
$|5x + 2| = 17$		$|5x + 2| = -7$

Out of these two new equations that we have obtained, one of them has no solution. The equation $|5x + 2| = -7$ has no solution since an absolute value cannot be negative. Hence we discard $|5x + 2| = -7$. However, the other equation which is $|5x + 2| = 17$ can be solved. Hence, let us solve it by splitting it into two equations as follows:

$5x + 2 = 17$ or $5x + 2 = -17$

$5x = 17 - 2$ $5x = -17 - 2$

$5x = 15$ $5x = -19$

$x = \dfrac{15}{5}$ $x = \dfrac{-19}{5}$

$x = 3$ $x = -3\dfrac{4}{5}$

Therefore, $x = 3$ or $x = -3\dfrac{4}{5}$

b. $||7 - 2x| - 8| = 5$

Starting with the outermost bars, we split the overall equation into two equations as follows:

$|7 - 2x| - 8 = 5$ or $|7 - 2x| - 8 = -5$

$|7 - 2x| = 5 + 8$ $|7 - 2x| = -5 + 8$

$|7 - 2x| = 13$ ....Equation (1) $|7 - 2x| = 3$ ....Equation (2)

Now we have to take each of the equations above and split it into two equations. Let us split equation (1) into two equations as follows:

$7 - 2x = 13$ or $7 - 2x = -13$

$-2x = 13 - 7$ $-2x = -13 - 7$

$-2x = 6$ $-2x = -20$

$x = \dfrac{6}{-2}$ $x = \dfrac{-20}{-2}$

$x = -3$ $x = 10$

Hence $x = -3$ or $x = 10$

Let us take equation (2) above and split it into two equations as follows:

$7 - 2x = 3$ or $7 - 2x = -3$

$-2x = 3 - 7$ $-2x = -3 - 7$

$-2x = -4$ $-2x = -10$

$x = \dfrac{-4}{-2}$ $x = \dfrac{-10}{-2}$

$x = 2$ $x = 5$

Hence $x = 2$ or $x = 5$

Therefore the overall solutions to the original equation are: $x = -3, 10, 2$ or 5

7. Solve the following equations:

a. $|2x-3| = |5x+1| + 1$

b. $|x+7| - |3x-1| = -4$

Solution

a. $|2x-3| = |5x+1| + 1$

There are three possible equations that we are going to solve.

(i) The first equation that we are going to solve is obtained by simply dropping the absolute value bars in the equation. This gives:

$$2x-3 = 5x+1+1$$
$$2x-5x = 1+1+3$$
$$-3x = 5$$
$$x = -\frac{5}{3}$$

(ii) The second equation that we are going to solve is obtained by making the two absolute value terms to be negative. This gives:

$$-(2x-3) = -(5x+1) + 1$$
$$-2x+3 = -5x-1+1$$
$$-2x+5x = -1+1-3$$
$$3x = -3$$
$$x = \frac{-3}{3}$$
$$x = -1$$

(iii) Lastly, the third equation that we have to solve is obtained by making only the absolute value term on the left hand side to be negative. This gives:

$$-(2x-3) = 5x+1+1$$
$$-2x+3 = 5x+1+1$$
$$-2x-5x = 1+1-3$$
$$-7x = -1$$
$$x = \frac{-1}{-7}$$
$$x = \frac{1}{7}$$

Hence the three values of x obtained are $-\frac{5}{3}, -1$ and $\frac{1}{7}$.

Now we have to check which of these values are solutions of the equation. We do this by substituting each of the value of x into the equation. The equation is:

$$|2x-3| = |5x+1| + 1$$

When $x = -\frac{5}{3}$, we have:

$$|2(-\tfrac{5}{3}) - 3| = |5(-\tfrac{5}{3}) + 1| + 1$$

$$|-\tfrac{10}{3} - 3| = |-\tfrac{25}{3} + 1| + 1$$

$$|\tfrac{-10-9}{3}| = |\tfrac{-25+3}{3}| + 1$$

$$|\tfrac{-19}{3}| = |\tfrac{-22}{3}| + 1$$

$$\tfrac{19}{3} = \tfrac{22}{3} + 1 \qquad \text{(Take note of the removal of the negative sign when the bars are removed}$$

$$\tfrac{19}{3} = \tfrac{22+3}{3}$$

$$\tfrac{19}{3} = \tfrac{25}{3}$$

Since both sides of the equation are not equal, then $x = -\dfrac{5}{3}$ is not a solution of the equation.

When $x = -1$, we have:

$$|2x - 3| = |5x + 1| + 1$$

$$|2(-1) - 3| = |5(-1) + 1| + 1$$

$$|-2 - 3| = |-5 + 1| + 1$$

$$|-5| = |-4| + 1$$

$$5 = 4 + 1$$

$$5 = 5$$

Since both sides are equal, then $x = -1$ is a solution of the equation.

When $x = \dfrac{1}{7}$, we have:

$$|2(\tfrac{1}{7}) - 3| = |5(\tfrac{1}{7}) + 1| + 1$$

$$|\tfrac{2}{7} - 3| = |\tfrac{5}{7} + 1| + 1$$

$$|\tfrac{2-21}{7}| = |\tfrac{5+7}{7}| + 1$$

$$|\tfrac{-19}{7}| = |\tfrac{12}{7}| + 1$$

$$\tfrac{19}{7} = \tfrac{12}{7} + 1$$

$$\tfrac{19}{7} = \tfrac{12+7}{7}$$

$$\tfrac{19}{7} = \tfrac{19}{7}$$

Since both sides are equal, then $x = \dfrac{1}{7}$ is a solution of the equation.

Therefore the solutions of the equation are $x = -1$ and $x = \dfrac{1}{7}$

b. $|x+7| - |3x-1| = -4$

Let us rearrange the equation in order to have only one of the absolute value expression on one side of the equation. This gives:

$|x+7| = |3x-1| - 4$

Let us solve each of the three possible equations from this equation.

(i) The first equation that we are going to solve is obtained by simply dropping the absolute value bars in the equation. This gives:

$x+7 = 3x-1-4$

$x-3x = -1-4-7$

$-2x = -12$

$x = \dfrac{-12}{-2}$

$x = 6$

(ii) The second equation is obtained by making the two absolute value terms to be negative. This gives:

$-(x+7) = -(3x-1) - 4$

$-x-7 = -3x+1-4$

$-x+3x = 1-4+7$

$2x = 4$

$x = \dfrac{4}{2}$

$x = 2$

(iii) The third equation that we have to solve is obtained by making only the absolute value term on the left hand side to be negative. This gives:

$-(x+7) = 3x-1-4$

$-x-7 = 3x-1-4$

$-x-3x = -1-4+7$

$-4x = 2$

$x = \dfrac{2}{-4}$

$x = -\dfrac{1}{2}$

Hence the three values of x obtained are 6, 2 and $-\dfrac{1}{2}$

Now, we have to check which of these values are solutions of the equation. We do this by substituting each of the values of x into the equation. The equation is:

$|x+7| = |3x-1| - 4$

When $x = 6$, we have:

$|6+7| = |3(6)-1| - 4$

$|13| = |18 - 1| - 4$

$13 = |17| - 4$

$13 = 17 - 4$

$13 = 13$

Since both sides of the equation are equal, then $x = 6$ is a solution of the equation.

When $x = 2$, we have:

$|2 + 7| = |3(2) - 1| - 4$

$|9| = |6 - 1| - 4$

$9 = |5| - 4$

$9 = 5 - 4$

$9 = 1$

Since both sides of the equation are not equal, then $x = 2$ is not a solution of the equation.

When $x = -\dfrac{1}{2}$, we have:

$|-\dfrac{1}{2} + 7| = |3(-\dfrac{1}{2}) - 1| - 4$

$|\dfrac{-1+14}{2}| = |-\dfrac{3}{2} - 1| - 4$

$|\dfrac{13}{2}| = |\dfrac{-3-2}{2}| - 4$

$\dfrac{13}{2} = |\dfrac{-5}{2}| - 4$

$\dfrac{13}{2} = \dfrac{5}{2} - 4$

$\dfrac{13}{2} = \dfrac{5 - 8}{2}$

$\dfrac{13}{2} = -\dfrac{3}{2}$

Since both sides of the equation are not equal, then $x = -\dfrac{1}{2}$ is not a solution of the equation.

Therefore the only solution of the equation is $x = 6$

8. Solve the following equations:

a. $-2|4x + 9| = -22$

b. $3|x - 2| + 5 = -2|x - 2| + 10$

Solution

a. $-2|4x + 9| = -22$

Divide both sides by −2. This gives:

$|4x + 9| = \dfrac{-22}{-2}$

$|4x + 9| = 11$

We now split this equation into two equations as follows:

$$4x + 9 = 11 \qquad \text{or} \qquad 4x + 9 = -11$$

$$4x = 11 - 9 \qquad\qquad\qquad 4x = -11 - 9$$

$$4x = 2 \qquad\qquad\qquad\qquad 4x = -20$$

$$x = \frac{2}{4} \qquad\qquad\qquad\qquad x = \frac{-20}{4}$$

$$x = \frac{1}{2} \qquad\qquad\qquad\qquad x = -5$$

Therefore, $x = \dfrac{1}{2}$ or $x = -5$

b. $3|x - 2| + 5 = -2|x - 2| + 10$

This equation is similar to 3m + 5 = −2m + 10, where m is $|x - 2|$. Hence we can solve for m. However, we are not going to use m, we will solve the equation using $|x - 2|$ as a variable. This is done as follows.

$$3|x - 2| + 5 = -2|x - 2| + 10$$

Collect terms in $|x - 2|$ on the left hand side of the equation. This gives:

$3|x - 2| + 2|x - 2| = 10 - 5$ (The left hand side is similar to 3m + 2m which will give 5m)

$5|x - 2| = 5$ (The left side is like 5m which is $5|x - 2|$ if we imagine m = $|x - 2|$)

Divide both sides by 5. This gives:

$$|x - 2| = \frac{5}{5}$$

$$|x - 2| = 1$$

We now split this equation into two equations as follows:

$$x - 2 = 1 \qquad \text{or} \qquad x - 2 = -1$$

$$x = 1 + 2 \qquad\qquad\qquad x = -1 + 2$$

$$x = 3 \qquad\qquad\qquad\qquad x = 1$$

Therefore, $x = 3$ or $x = 1$

9. Solve the following equations:

a. $|2x^2 - 7x + 6| = 0$

b. $|x^2 + 5x - 80| = 4$

<u>Solutions</u>

a. $|2x^2 - 7x + 6| = 0$

Since the right hand side of the equation is zero, we can split this equation to only one equation. This gives:

$$2x^2 - 7x + 6 = 0$$

Solving this quadratic equation by factorization gives:

$$2x^2 - 4x - 3x + 6 = 0$$

$$2x(x-2) - 3(x-2) = 0$$
$$(x-2)(2x-3) = 0$$

Hence, $x = 2$ or $\dfrac{3}{2}$.   (After equating each bracket above to zero and solving for x)

b. $|x^2 + 5x - 80| = 4$

We can split this equation into two equations as follows:

$x^2 + 5x - 80 = 4$ ....Equation (1)

$x^2 + 5x - 80 = -4$ ....Equation (2)

Let us solve equation (1) as follows:

$$x^2 + 5x - 80 = 4$$
$$x^2 + 5x - 80 - 4 = 0$$
$$x^2 + 5x - 84 = 0$$

Solving this equation by factorization gives:

$$(x\quad)(x\quad) = 0$$

We now find two numbers such that their product is −84 and their sum is +5. The two numbers are +12 and −7. The two number are entered into the brackets above to give:

$$(x + 12)(x - 7) = 0$$

Hence, $x = -12$ or $x = 7$

Let us now solve equation (2) above as follows:

$$x^2 + 5x - 80 = -4$$
$$x^2 + 5x - 80 + 4 = 0$$
$$x^2 + 5x - 76 = 0$$

Let us use quadratic formula to solve this equation. From the equation, a = 1, b = 5 and c = −76

$$x = \frac{-b \pm \sqrt{b^2 - 4ac}}{2a}$$

$$= \frac{-5 \pm \sqrt{5^2 - [4(1)(-76)]}}{2(1)}$$

$$= \frac{-5 \pm \sqrt{25 - (-304)}}{2}$$

$$= \frac{-5 \pm \sqrt{25 + 304}}{2}$$

$$= \frac{-5 \pm \sqrt{329}}{2}$$

$$= \frac{-5 \pm 18.14}{2}$$

$$x = \frac{-5 + 18.14}{2} \quad \text{or} \quad x = \frac{-5 - 18.14}{2}$$

$$= \frac{13.14}{2} \quad \text{or} \quad \frac{-23.14}{2}$$

$$x = 6.57 \quad \text{or} \quad x = -11.57$$

Therefore, the four solutions of the equation are $x = -12, 7, 6.57$ and -11.57.

Exercise 4

1. Solve $|x| = 9$
2. Solve the following equations:
a. $|5x + 7| = 17$
b. $|3x + 1| = 8$
c. $|15 - 4x| = 18$

3. Solve the following equation:
a. $|7x + 13| = 8$
b. $|5x - 27| = -14$

4. Solve the following equations:
a. $|4x - 3| + 6 = 17$
b. $|x + 9| - 11 = -7$

5. Solve the following absolute value equations:
a. $|3x - 6| = x - 2$
b. $|8x + 5| = 7x + 40$
c. $|2x - 5| = |10x - 1|$

6. Solve the following absolute value equations:
a. $||2x + 7| - 5| = 6$
b. $||9 - 4x| - 8| = 1$

7. Solve the following equations:
a. $|3x - 1| = |2x + 5| + 6$
b. $|5x + 2| - |11x - 3| = -1$

8. Solve the following equations:
   a. $-5|4x + 9| = 25$
   b. $2|2x - 3| + 1 = -|2x - 3| + 22$

9. Solve the following equations:
   a. $|3x^2 - 7x - 6| = 0$
   b. $|x^2 + 14x - 68| = 4$

10. Solve the following equations:
   a. $|5x^2 - 13x + 6| = 0$
   b. $|2x^2 + 9x - 12| = 3$

CHAPTER 5
INEQUALITIES INVOLVING ABSOLUTE VALUES, QUOTIENT AND SQUARE FUNCTIONS

The modulus or absolute value of a number is the size of the number without its sign. It is denoted by a vertical line enclosing a number. For example $|5| = 5$ and $|-5| = 5$. A negative number enclosed in the vertical lines is regarded as positive.

If an inequality is given by : $|x| < 2$, it means that x is in the range: $-2 < x < 2$, i.e. $x < 2$ and $x > -2$

Examples

1. Solve the inequality: $|5x - 2| < 13$

<u>Solution</u>

$|5x - 2| < 13$

Note that this also means $|-(5x - 2)| < 13$.

The range of the term in modulus is given by:

$-13 < 5x - 2 < 13$

We now take them separately and solve. Taking the first part (left hand part) gives:

$-13 < 5x - 2$

$-13 + 2 < 5x$

$-11 < 5x$

$-\dfrac{11}{5} < x$

Or $x > -\dfrac{11}{5}$ (Make sure the open side of the inequality sign still faces x when rearranging it).

Let us now take the other part of the inequality range and solve. This gives:

$5x - 2 < 13$

$5x < 13 + 2$

$5x < 15$

$x < \dfrac{15}{5}$

$x < 3$

We now combine the two results to obtain the range of the solution as follows:

$-\dfrac{11}{5} < x < 3$

Note that the opened end of the inequality sign is facing x in the solution $x > -\dfrac{11}{5}$, while the closed end (elbow end) of the inequality sign is facing x in the solution $x < 3$. Therefore, when combining the solution (i.e. $-\dfrac{11}{5} < x < 3$) you have to ensure that the right part of the inequality sign is facing x.

2. Solve the inequality $|3x + 7| \geq 10$

<u>Solution</u>

$\qquad |3x + 7| \geq 10$

This means: $(3x + 7) \geq 10$. This directly gives: $3x + 7 \geq 10$

It also means $-(3x + 7) \geq 10$. When we divide both sides by -1, it gives

$$\dfrac{-(3x + 7)}{-1} \geq \dfrac{10}{-1}$$

$\therefore \quad 3x + 7 \leq -10$

Take note of the reversal of the inequality sign since both sides of the inequality were divided by a negative number.

Hence, the two inequalities obtained from $|3x + 7| \geq 10$ are:

$\qquad 3x + 7 \geq 10$ and $3x + 7 \leq -10$

We now take each inequality and solve. This gives:

$\qquad 3x + 7 \geq 10$

$\qquad 3x \geq 10 - 7$

$\qquad 3x \geq 3$

$\therefore \quad x \geq \dfrac{3}{3}$

$\qquad x \geq 1$

Solving the second inequality obtained gives:

$\qquad 3x + 7 \leq -10$

$\qquad 3x \leq -10 - 7$

$\qquad 3x \leq -17$

$\qquad x \leq -\dfrac{17}{3}$

$\qquad x \leq -5\dfrac{2}{3}$

$\therefore \quad |3x + 7| \geq 10$ is satisfied when $x \geq 1$ and $x \leq -5\dfrac{2}{3}$

Note that these two results cannot be combined together since the solutions do not range from one to the other. This is because $x \geq 1$ means x can be 1, 2, 3 ..., while $x \leq -5\dfrac{2}{3}$ means that x can be $-5\dfrac{2}{3}$, -6, -7.... Hence, the two set of values do not meet and so cannot be combined together. A solution such as $x \leq 1$ and $x \geq -3$ can be combined together to give $-3 \leq x \leq 1$. This is because the values meet as follows: $-3, -2, -1, 0, 1$. Therefore, take note of this point before combining any pair of solutions.

3. Solve: $|4x - 9| \leq 3$

Solution

$\quad |4x - 9| \leq 3$

The two inequalities that can be obtained from this are:

$\quad 4x - 9 \leq 3$

And $\quad 4x - 9 \geq -3$ (This is obtained by simply reversing the inequality sign and changing the positive number on the right to be negative)

We now solve each of the inequality as follows:

$\quad 4x - 9 \leq 3$

$\quad 4x \leq 3 + 9$

$\quad 4x \leq 12$

$\quad x \leq \dfrac{12}{4}$

$\quad x \leq 3$

And: $\quad 4x - 9 \geq -3$

$\quad 4x \geq -3 + 9$

$\quad 4x \geq \dfrac{6}{4}$

$\quad x \geq \dfrac{3}{2}$

Wee can now combine the solutions to obtain: $\dfrac{3}{2} \leq x \leq 3$

$\therefore$ The inequality $|4x - 9| \leq 3$ is satisfied within the range $\dfrac{3}{2} \leq x \leq 3$

4. Solve the inequality $|3m - 5| > 4$

Solution

$\quad |3m - 5| > 4$

The two inequalities that can be obtained from this are:

$\quad 3m - 5 > 4$

And $\quad 3m - 5 < -4$ (Simply reverse the inequality sign and assign a negative sign to the number on the right hand side)

We now solve each of the inequality as follows:

$\quad 3m - 5 > 4$

$\quad 3m > 4 + 5$

$\quad 3m > 9$

$\quad m > \dfrac{9}{3}$

$\quad m > 3$

And $\quad 3m - 5 < -4$

$\quad 3m < -4 + 5$

$$3m < 1$$

$$m < \frac{1}{3}$$

∴ The inequality $|3m - 5| > 4$ is satisfied when m > 3 and m < $\frac{1}{3}$ (They cannot be combined)

Inequalities Involving Quotients

If $\frac{x}{y} > 0$, then $\frac{x}{y}$ has to be a positive value.

For $\frac{x}{y}$ to be positive either both x and y must be positive or both x and y must be negative.

If $\frac{x}{y} < 0$, then $\frac{x}{y}$ has to be a negative value. For $\frac{x}{y}$ to be negative, only x or only y must be negative.

However, for inequalities such as $\frac{x}{y} \geq 0$ or $\frac{x}{y} \leq 0$, the same rules above applies to the numerator (i.e. x), but y ≥ 0 or y ≤ 0 cannot be used for the denominator so that it will not render the fraction undefined. Hence we use y > 0 or y < 0 for the denominator. This is because a fraction such as $\frac{2}{0}$ is undefined. So, in $\frac{x}{y} \geq 0$ or $\frac{x}{y} \leq 0$, y ≥ 0 or y ≤ 0 is not possible since it will make the fraction undefined.

Examples

1. Solve the inequality $\frac{x-2}{3x+9} > 0$

Solution

$$\frac{x-2}{3x+9} > 0$$

For the inequality above to be greater than zero, (i.e. > 0) it means that it has to be positive. For it to be positive, either the numerator and denominator are both positive (i.e. > 0) or they are both negative (i.e. < 0). This means that:

a. When both numerator and denominator are positive, then:

$$x - 2 > 0$$

and $3x + 9 > 0$

Solving each of these gives:

$$x - 2 > 0$$

∴ $x > 2$

and $3x + 9 > 0$

$$3x > -9$$

$$x > -\frac{9}{3}$$

$$x > -3$$

Therefore, we now have two possible solutions. But we have to test each of the solution to see

48

if they are true. This is done by substituting a possible value in each solution into the original inequality.

Let us test the solution $x > 2$. A possible value here is 3 (since $3 > 2$). Now substitute 3 for x in the original inequality. This gives:

$$\frac{x-2}{3x+9} > 0$$

$$\frac{3-2}{3(3)+9} > 0$$

$$\frac{1}{9+9} > 0$$

$$\frac{1}{18} > 0 \quad \text{(This is true)}$$

Hence, the solution $x > 2$ is correct.

Let us test the solution $x > -3$. A possible value here is -2 (Since $-2 > -3$). Now substitute -2 for x in the original inequality. This gives:

$$\frac{x-2}{3x+9} > 0$$

$$\frac{-2-2}{3(-2)+9} > 0$$

$$\frac{-4}{-6+9} > 0$$

$$\frac{-4}{3} > 0$$

$$-\frac{4}{3} > 0 \quad \text{(This is not true since 0 is greater)}$$

Hence, the solution $x > -3$ is not correct.

b. Now we are through with when numerator and denominator are positive. When numerator and denominator are negative, then:

$$x - 2 < 0$$

And $\quad 3x + 9 < 0$

Solving each of them gives:

$$x - 2 < 0$$

$$x < 2$$

And $\quad 3x + 9 < 0$

$$3x < -9$$

$$x < -\frac{9}{3}$$

$$x < -3$$

Hence, $x < 2$ or $x < -3$

Let us now test each of the solution to see if they are true.

When $x < 2$, a possible value here is 1 (since $1 < 2$). Now substitute 1 for x in the inequality. This gives:

$$\frac{x-2}{3x+9} > 0$$

$$\frac{1-2}{3(1)+9} > 0$$

$$\frac{-1}{3+9} > 0$$

$$-\frac{1}{12} > 0 \quad \text{(This is not true since 0 is greater than } -\frac{1}{12})$$

Hence, the solution $x < 2$ is not correct.

Similarly, when $x < -3$, a possible value here is –4 (since –4 < –3). We now substitute –4 for x in the inequality. This gives:

$$\frac{x-2}{3x+9} > 0$$

$$\frac{-4-2}{3(-4)+9} > 0$$

$$\frac{-6}{-12+9} > 0$$

$$\frac{-6}{-3} > 0$$

$$2 > 0 \quad \text{(This is true)}$$

Hence, the solution $x < -3$ is correct.

Hence, in all our four solutions, the two that are correct are $x > 2$ and $x < -3$

Therefore, $\frac{x-2}{3x+9} > 0$ is satisfied when $x > 2$ and $x < -3$

2. Solve the inequality $\frac{2m+6}{m-4} < 0$

Solution

$$\frac{2m+6}{m-4} < 0$$

Example (1) above appears complex because I took my time to thoroughly explain it. In order to avoid making this example as long as example (1), I will use a more direct procedure.

Here, let us solve each of the numerators and denominators using the equality sign as follows:

$$2m + 6 = 0$$

$$2m = -6$$

$$m = -\frac{6}{2}$$

$$m = -3$$

Applying inequality to the solution above implies that:

$$m < -3 \text{ or } m > -3 \quad \text{(Use the two inequality signs)}$$

Let us test each of them to find the correct solution.

When m < –3, a possible value here is – 3.5 (since –3.5 < –3. Testing with a value that just begins the solution puts us on a safer side). Now, substitute –3.5 into the inequality. This gives:

$$\frac{2m + 6}{m - 4} < 0$$

$$\frac{2(-3.5) + 6}{-3.5 - 4} < 0$$

$$\frac{-7 + 6}{-3.5 - 4} < 0$$

$$\frac{-1}{-7.5} < 0$$

$$\frac{1}{7.5} < 0 \qquad \text{(This is not true)}$$

When m > –3, a possible value here is –2.5. Substituting this value into the inequalities gives:

$$\frac{2m + 6}{m - 4} < 0$$

$$\frac{2(-2.5) + 6}{-2.5 - 4} < 0$$

$$\frac{-5 + 6}{-2.5 - 4} < 0$$

$$\frac{1}{-6.5} < 0 \qquad \text{(This is true)}$$

Hence m > –3 is correct.

Let us now carry out the same procedure with the denominator.

$$m - 4 = 0$$

$$\therefore \qquad m = 4$$

Hence, m < 4 or m > 4   (When the two inequality signs are applied)

When m < 4, let us substitute 3.5 into the inequality as follows:

$$\frac{2m + 6}{m - 4} < 0$$

$$\frac{2(3.5) + 6}{3.5 - 4} < 0$$

$$\frac{7 + 6}{3.5 - 4} < 0$$

$$\frac{13}{-0.5} < 0$$

$$-26 < 0 \qquad \text{(This is true)}$$

Hence, m < 4 is correct

Since this is correct, it means that the other solution of m > 4 is wrong since we already have our two possible solutions. However, let us test the other value.

When m > 4, let us substitute 4.5 into the inequality as follows:

$$\frac{2m + 6}{m - 4} < 0$$

$$\frac{2(4.5) + 6}{4.5 - 4} < 0$$

$$\frac{9+6}{0.5} < 0$$

$$\frac{15}{0.5} < 0$$

$$30 < 0 \qquad \text{(This is not true)}$$

Therefore, our two possible solutions are m > –3 or m < 4. They can be combined to give:

$$-3 < x < 4$$

Therefore, $\frac{2m+6}{m-4} < 0$, is satisfied when –3 < x < 4

3. Solve the inequality $\frac{2x-3}{x+2} \leq 1$

<u>Solution</u>

$$\frac{2x-3}{x+2} \leq 1$$

Let us first make the right hand side to be zero.

$$\therefore \quad \frac{2x-3}{x+2} - 1 \leq 0$$

Simplifying with x + 2 as the LCM gives:

$$\frac{2x-3-1(x+2)}{x+2} \leq 0$$

$$\frac{2x-3-x-2}{x+2} \leq 0$$

$$\frac{2x-x-3-2}{x+2} \leq 0$$

$$\frac{x-5}{x+2} \leq 0$$

At this point, we can now take the numerator and solve for x as follows:

$$x - 5 = 0$$

$$x = 5$$

Applying inequality to this solution gives:

$$x \leq 5 \text{ or } x \geq 5$$

If we test the inequality when x = 4.5 (for $x \leq 5$) and x = 5.5 (for $x \geq 5$) respectively, we will find out that the true solution is $x \leq 5$

Taking the denominator gives:

$$x + 2 = 0$$

$$x = -2$$

Applying inequality to this solution gives:

$$x < -2 \text{ or } x > -2$$

Note that we cannot use $\leq$ or $\geq$ for the denominator since the denominator cannot be zero. If this is done, it will make the denominator undefined since fractions such as $\frac{5}{0}$ is undefined.

Now, back to $x < -2$ or $x > -2$. If we test the inequality with $x = -2.5$ (for $x < -2$) and $x = -1.5$ (for $x > -2$), respectively, we will find out that the true solution is $x > -2$.

$\therefore$ The solutions are $x \leq 5$ or $x > -2$

Hence, $\dfrac{2x-3}{x+2} \leq 1$ is satisfied when $-2 < x \leq 5$

4. Solve the inequality $\dfrac{2-b}{b+3} \geq 4$

Solution

$$\dfrac{2-b}{b+3} \geq 4$$

Let us first make the right hand side to be zero.

$$\dfrac{2-b}{b+3} \geq 4$$

$$\dfrac{2-b}{b+3} - 4 \geq 0$$

$$\dfrac{2-b-4(b+3)}{b+3} \geq 0$$

$$\dfrac{2-b-4b-12}{b+3} \geq 0$$

$$\dfrac{-5b-10}{b+3} \geq 0$$

Taking the numerator gives:

$$-5b - 10 = 0$$

$$-5b = 10$$

$$b = \dfrac{10}{-5}$$

$$b = -2$$

$\therefore$ $b \geq -2$ or $b \leq -2$

Testing possible values (-1.5 and -2.5) in each of these two solutions shows that $b \leq -2$ is the correct solution.

Taking the denominator gives:

$$b + 3 = 0$$

$$b = -3$$

$\therefore$ $b < -3$ or $b > -3$ (Note that $\geq$ or $\leq$ cannot be used for the denominator)

Testing values (-3.5 and -2.5) from each of the solution above shows that $b > -3$ is the correct solution.

Hence the solution are $b \leq -2$ or $b > -3$

Therefore, $\dfrac{2-b}{b+3} \geq 4$ is satisfied when $-3 < b \leq -2$

Inequalities Involving Square Functions

If $x^2 < c$, then $x < \sqrt{c}$ or $x > -\sqrt{c}$

Similarly, if $x^2 > c$, then $x > \sqrt{c}$ or $x < -\sqrt{c}$

These rules are applied in inequalities involving square functions.

Examples

1. Solve the inequality $x^2 > 16$

<u>Solution</u>

$\quad x^2 > 16$

$\therefore \quad x > \sqrt{16}$

$\quad x > 4$

Or $\quad x < -\sqrt{16}$ (Reverse the inequality sign and introduce a negative sign)

$\quad x < -4$

$\therefore \quad x > 4$ or $x < -4$

Therefore $x^2 > 16$ is satisfied when $x > 4$ or $x < -4$

2. Solve the inequality: $y^2 < 25$

<u>Solution</u>

$\quad y^2 < 25$

$\therefore \quad y < \sqrt{25}$

$\quad y < 5$

Or $\quad y > -\sqrt{25}$ (Reverse the inequality sign and introduce a negative sign)

$\quad y > -5$

Therefore $y^2 < 25$ is satisfied when $-5 < y < 5$

3. Solve: $(2a - 3)^2 \geq 49$

<u>Solution</u>

$\quad (2a - 3)^2 \geq 49$

$\quad 2a - 3 \geq \sqrt{49}$

$\quad 2a - 3 \geq 7$

$\quad 2a \geq 7 + 3$

$\quad 2a \geq 10$

$\quad a \geq \dfrac{10}{2}$

$\quad a \geq 5$

Or $\quad (2a - 3) \leq -\sqrt{49}$ (Reverse the inequality sign and introduce a negative sign)

$\quad 2a - 3 \leq -7$

$$2a \leq -7 + 3$$
$$2a \leq -4$$
$$a \leq -\frac{4}{2}$$
$$a \leq -2$$

Therefore $(2a - 3)^2 \geq 49$ is satisfied when $x \geq 5$ or $x \leq -2$

4. Solve: $7 - 4y^2 \leq -29$

Solution

$$7 - 4y^2 \leq -29$$
$$-4y^2 \leq -29 - 7$$
$$-4y^2 \leq -36$$
$$y^2 \geq \frac{-36}{-4} \quad \text{(Take note of the reversal of the inequality sign due to division by a negative}$$

number)
$$y^2 \geq 9$$

Hence, $y \geq \sqrt{9}$
$$y \geq 3$$
Or $y \leq -\sqrt{9}$
$$y \leq -3$$
$\therefore$ $7 - 4y^2 \leq -29$ is satisfied when $y \geq 3$ or $y \leq -3$

5. Solve $5m^2 + 2 \geq 12$

Solution

$$5m^2 + 2 \geq 12$$
$$5m^2 \geq 12 - 2$$
$$5m^2 \geq 10$$
$$m^2 \geq \frac{10}{5}$$
$$m^2 \geq 2$$
$\therefore$ $m \geq \sqrt{2}$
Or $m \leq -\sqrt{2}$

Therefore $5m^2 + 2 \geq 12$ is satisfied when $m \geq \sqrt{2}$ or $m \leq -\sqrt{2}$

Exercise 5

Solve the following inequalities:

1. $|5x + 3| \geq 7$

2. $|3x - 2| < 17$

3. $|x - 8| \leq 11$

4. $|7m + 5| > 2$

5. $\dfrac{x - 5}{2x + 6} > 0$

6. $\dfrac{3m + 9}{m - 2} < 0$

7. $\dfrac{2x - 7}{x - 2} \leq 1$

8. $\dfrac{9 - x}{2x + 4} \geq 5$

9. $2x^2 > 18$

10. $y^2 < 36$

11. $(3a - 5)^2 \geq 125$

12. $5 - 2y^2 \leq -13$

13. $9m^2 - 1 \geq 8$

14. $\dfrac{19 - 2x}{2x + 1} \geq 3$

15. $7 - 5y^2 \leq -73$

CHAPTER 6
INDICIAL EQUATIONS

An indicial equation is an equation in which the unknown variable is a power (index) in the equation. Usually the knowledge of indices and logarithms is applied in solving this type of equation.

Examples

1. Solve: $3^{3x+4} = 27^{2x+5}$

<u>Solution</u>

$$3^{3x+4} = 27^{2x+5}$$

In order to solve indicial equation, express both sides of the equation in the same base, and then equate the powers. The left hand side of the equation above has a base of 3. It is clear that the right hand side can also be expressed as a base of 3 because $27 = 3^3$. Hence, the equation now simplifies as follows:

$$3^{3x+4} = (3^3)^{2x-5}$$
$$3^{3x+4} = 3^{3(2x-5)} \quad \text{(From indices, } (a^x)^y = a^{xy})$$
$$3^{3x+4} = 3^{6x-15}$$

Since the bases on both sides of the equation are equal, it implies that the powers are also equal. Hence we ignore the bases and equate their powers as follows:

$$3x + 4 = 6x - 15$$
$$4 + 15 = 6x - 3x$$
$$19 = 3x$$
$$\therefore \quad x = \frac{19}{3}$$
$$x = 6\frac{1}{3}$$

2. Solve the equation: $\dfrac{8^{x+3}}{4^{3x-1}} = 32^{2x+7}$

<u>Solution</u>

$$\frac{8^{x+3}}{4^{3x-1}} = 32^{2x+7}$$

We can express each of the bases in the equation as a base of two. This gives:

$$\frac{(2^3)^{x+3}}{(2^2)^{3x-1}} = (2^5)^{2x+7} \quad \text{(Note that } 8 = 2^3, 4 = 2^2 \text{ and } 32 = 2^5)$$

$$\frac{2^{3(x+3)}}{2^{2(3x-1)}} = 2^{5(2x+7)} \quad \text{(Since } (a^x)^y = a^{xy})$$

$$\frac{2^{3x+9}}{2^{6x-2}} = 2^{10x+35}$$

$$\therefore \quad 2^{3x+9-(6x-2)} = 2^{10x+35} \quad \text{(From indices, } \frac{a^x}{a^y} = a^{x-y}, \text{ i.e. subtraction of powers)}$$

$$2^{3x+9-6x+2} = 2^{10x+35}$$

$$2^{-3x+11} = 2^{10x+35}$$

Since the bases are equal, it also means that the powers are equal. Hence, we equate the powers as follows:

$$-3x + 11 = 10x + 35$$

$$-3x - 10x = 35 - 11$$

$$-13x = 24$$

$$\therefore \quad x = -\frac{24}{13}$$

$$x = -1\frac{11}{13}$$

3. Solve: $\dfrac{125^{x-5}}{625^{2x}} = \dfrac{5^{3x+1}}{256^{6x-1}}$

<u>Solution</u>

$$\frac{125^{x-5}}{625^{2x}} = \frac{5^{3x+1}}{256^{6x-1}}$$

Let us express each base above as a base of 5 as follows:

$$\frac{(5^3)^{x-5}}{(5^4)^{2x}} = \frac{5^{3x+1}}{(5^2)^{6x-1}}$$

$$\frac{5^{3(x-5)}}{5^{4(2x)}} = \frac{5^{3x+1}}{5^{2(6x-1)}}$$

$$\frac{5^{3x-15)}}{5^{8x}} = \frac{5^{3x+1}}{5^{12x-2}}$$

Cross multiply to obtain:

$$5^{3x-15} \times 5^{12x-2} = 5^{3x+1} \times 5^{8x}$$

$$5^{3x-15+12x-2} = 5^{3x+1+8x} \quad \text{(From indices, } a^x \times a^y = a^{x+y} \text{ i.e. addition of powers)}$$

$$5^{15x-17} = 5^{11x+1}$$

We now equate the powers since the bases are equal.

$$\therefore \quad 15x - 17 = 11x + 1$$

$$15x - 11x = 1 + 17$$

$$4x = 18$$

$$x = \frac{18}{4}$$

$$= \frac{9}{2}$$

$$\therefore \quad x = 4\frac{1}{2}$$

4. Solve the equation $3^{2x} = 12$

Solution

$\qquad 3^{2x} = 12$

A careful look at this equation shows that the two sides of the equation cannot be expressed in the same base. Hence, we take the logarithm of both sides of the equation as follows:

$\qquad 3^{2x} = 12$

$\qquad \text{Log } 3^{2x} = \log 12$

∴ $\quad 2x\log 3 = \log 12 \qquad$ (Note that from logarithm, $\log x^y = y\log x$)

$\qquad 2x\,(0.4771) = 1.0792 \qquad$ (From calculator, $\log 3 = 0.4771$ and $\log 12 = 1.0792$)

$\qquad 0.9542x = 1.0792$

∴ $\qquad x = \dfrac{1.0792}{0.9542}$

$\qquad x = 1.13$

5. Solve the equation $5^{4x-1} = 18^{2x+3}$

Solution

$\qquad 5^{4x-1} = 18^{2x+3}$

It is clear that 5 and 18 cannot be expressed in the same base in whole number. Hence, we take the logarithm of both sides of the equation as follows:

$\qquad 5^{4x-1} = 18^{2x+3}$

$\qquad \text{Log} 5^{4x-1} = \text{Log} 18^{2x+3}$

$\qquad (4x - 1)\log 5 = (2x + 3)\log 18$

$\qquad (4x - 1)(0.6990) = (2x + 3)(1.2553) \qquad$ (Note that $\log 5 = 0.6990$ and $\log 18 = 1.2553$)

$\qquad 2.796x - 0.6990 = 2.5106x + 3.7659$

$\qquad 2.796x - 2.5106x = 3.7659 + 0.669$

$\qquad 0.2854x = 4.4649$

∴ $\qquad x = \dfrac{4.4649}{0.2854}$

$\qquad x = 15.64$

6. Solve $3.2^{x} \times 7.5^{3x+2} = 4^{2x+5}$

Solution

$\qquad 3.2^{x} \times 7.53^{3x+2} = 4^{2x+5}$

Taking the logarithm of both sides gives:

$\qquad \text{Log}(3.2^{x} \times 7.53^{3x+2}) = \text{Log} 4^{2x+5}$

$\qquad \text{Log} 3.2^{x} + \text{Log} 7.53^{3x+2} = \text{Log} 4^{2x+5} \qquad$ (Note that from logarithm, $\text{Log}(AB) = \text{Log} A + \text{Log} B$)

$\qquad x\text{Log} 3.2 + (3x + 2)\text{Log} 7.53 = (2x + 5)\text{Log} 4$

$\qquad x(0.5051) + (3x + 2)(0.8573) = (2x + 5)(0.6021)$

$$0.5051x + 2.5719x + 1.7146 = 1.2042x + 3.0105$$
$$0.5051x + 2.5719x - 1.2042x = 3.0105 - 1.7146$$
$$1.8728x = 1.2959$$
$$x = \frac{1.2959}{1.8728}$$
$$x = 0.692$$

7. Solve the equation $2^{x^2} = 4^{3x-4}$

Solution

$$2^{x^2} = 4^{3x-4}$$

Expressing both sides in the same base (i.e. 2) gives:

$$2^{x^2} = (2^2)^{3x-4}$$
$$2^{x^2} = 2^{2(3x-4)}$$
$$2^{x^2} = 2^{6x-8}$$

Since the bases are equal, we now equate the powers as follows:

$$x^2 = 6x - 8$$
$$x^2 - 6x + 8 = 0$$

Solving this equation by factorization gives:

$$(x-4)(x-2) = 0$$

This gives: $x = 4$, or $x = 2$

8. Solve the equation $9^{x^2} = 3^{5x-3}$

Solution

$$9^{x^2} = 3^{5x-3}$$

Expressing both sides in the same base of 3 gives:

$$(3^2)^{x^2} = 3^{5x-3}$$
$$3^{2x^2} = 3^{5x-3}$$

Equating the powers gives:

$$2x^2 = 5x - 3$$
$$2x^2 - 5x + 3 = 0$$

Solving this equation by factorization gives:

$$2x^2 - 2x - 3x + 3 = 0$$
$$2x(x-1) - 3(x-1) = 0$$
$$(x-1)(2x-3) = 0$$
$$\therefore \quad x = 1 \text{ or } x = \frac{3}{2}$$

9. Solve the equation: $4^x - 3 \times 2^x + 2 = 0$

Solution

This equation is slightly different from others since it contains the addition/subtraction of three terms. The terms having x as their powers can be expressed in the same base as follows.

$(2^2)^x - 3(2^x) + 2 = 0$

$(2^x)^2 - 3(2^x) + 2 = 0$ (Note that $(2^2)^x = (2^x)^2$, since $2 \times x$ is the same as $x \times 2$)

Since $(2^x)^2$ means (2^x) raise to the power 2, it means that this equation can be expressed as a quadratic equation.

Let $2^x = y$ ....Equation (1)

Substitute y for 2^x in the simplified equation above as follows

$(2^x)^2 - 3(2^x) + 2 = 0$

$y^2 - 3y + 2 = 0$ (Since $2^x = y$)

Solving this equation by factorization gives:

$(y - 1)(y - 2) = 0$

Hence, y = 1 or 2

When y = 1, we substitute 1 for y in equation (1) in order to find x. This gives:

$2^x = y$ ....Equation (1)

$2^x = 1$

$2^x = 2^0$ (Note that $2^0 = 1$)

Equating the powers gives:

$x = 0$

When y = 2, we substitute 2 for y in equation (1) in order to find x. This gives:

$2^x = y$ ....Equation (1)

$2^x = 2$

$2^x = 2^1$ (Note that $2^1 = 2$)

Equating the powers gives:

$x = 1$

Hence, $x = 0$ or $x = 1$.

10. Solve the equation $2^{2x} - 2^{1+x} - 8 = 0$

Solution

$2^{2x} - 2^{1+x} - 8 = 0$

In this equation, 2^{1+x} can be expressed as $2^1 \times 2^x$, since from indices $a^x \times a^y = a^{x+y}$. This also means that $a^{x+y} = a^x \times a^y$. Hence substituting $2^1 \times 2^x$ for 2^{1+x} gives:

$2^{2x} - 2^{1+x} - 8 = 0$

$2^{2x} - 2^1 \times 2^x - 8 = 0$

$(2^x)^2 - 2(2^x) - 8 = 0$

Now, let $2^x = b$ ....Equation (1)

Substitute b for 2^x in the equation above. This gives:

$$(2^x)^2 - 2(2^x) - 8 = 0$$
$$b^2 - 2b - 8 = 0 \qquad \text{(Since } 2^x = b)$$

Solving this equation by factorization gives:

$$(b - 4)(b + 2) = 0$$
$$\therefore \qquad b = 4 \quad \text{or} \quad b = -2$$

When b = 4, we substitute 4 for b in equation (1). This gives:

$$2^x = b \text{.....................Equation (1)}$$
$$2^x = 4$$
$$\therefore \qquad 2^x = 2^2$$

Equating powers gives:

$$x = 2$$

Note that the other solution of b = −2 has been discarded since substituting −2 for b in equation (1) cannot be solved, i.e.:

$$2^x = -2 \text{ has no solution}$$

11. Solve: $3^{1+2x} + 2 \times 3^x - 1 = 0$

<u>Solution</u>

$$3^{1+2x} + 2 \times 3^x - 1 = 0$$
$$3^1 \times 3^{2x} + 2 \times 3^x - 1 = 0 \qquad \text{(Note that } 3^1 \times 3^{2x} = 3^{1+2x} \text{ from indices)}$$
$$3(3^x)^2 + 2(3^x) - 1 = 0$$

Let $3^x = y$ ....Equation (1)

Substitute y for 3^x in the equation above. This gives:

$$3(3^x)^2 + 2(3^x) - 1 = 0$$
$$3y^2 + 2y - 1 = 0 \qquad \text{(Since } 3^x = y)$$

Solving this equation by factorization gives:

$$3y^2 + 3y - y - 1 = 0$$
$$3y(y + 1) - 1(y + 1) = 0$$
$$(y + 1)(3y - 1) = 0$$
$$\therefore \qquad y = -1 \quad \text{or} \quad y = \frac{1}{3}$$

When $y = \frac{1}{3}$, we substitute $\frac{1}{3}$ for y in equation (1). This gives:

$$3^x = y \text{.....................Equation (1)}$$
$$3^x = \frac{1}{3}$$
$$\therefore \qquad 3^x = 3^{-1} \qquad \text{(Note that } 3^{-1} = \frac{1}{3})$$

Equating powers gives:

$$x = -1$$

Note that the other solution of y = −1 has been discarded since we cannot use it to solve for x.

12. Solve the simultaneous equation: $9^{2x-y} = 3$ and $16^{x+y} = 8$

Solution

Let us simplify each of the equation above one after the other.

$9^{2x-y} = 3$

Expressing both sides of the equation in the same base gives:

$(3^2)^{2x-y} = 3^1$

$3^{2(2x-y)} = 3^1$

$3^{4x-2y} = 3^1$

Equating the powers gives:

$4x - 2y = 1$ ....Equation (1)

Similarly, $16^{x+y} = 8$ can be expressed in base 2 as follows:

$(2^4)^{x+y} = 2^3$

$2^{4(x+y)} = 2^3$

$2^{4x+4y} = 3^3$

Equating the powers gives:

$4x + 4y = 3$ ....Equation (2)

Bring equation 1 and 2 together in order to solve them simultaneously. This gives:

$4x - 2y = 1$ ....Equation (1)

$\underline{4x + 4y = 3}$ ....Equation (2)

Equation (2) − (1): $6y = 2$ (Note that $4y - (-2y) = 6y$, and $3 - 1 = 2$)

$$y = \frac{2}{6}$$

$$y = \frac{1}{3}$$

Substitute $\frac{1}{3}$ for y in equation (1).

$4x - 2y = 1$ ....Equation (1)

$4x - 2(\frac{1}{3}) = 1$

$4x - \frac{2}{3} = 1$

$4x = 1 + \frac{2}{3}$

$4x = 1\frac{2}{3}$

$4x = \frac{5}{3}$

$x = \frac{\frac{5}{3}}{4}$

$= \frac{5}{3} \div 4$

$$= \frac{5}{3} \times \frac{1}{4} \qquad \text{(Note that 4 also means } \frac{4}{1}\text{)}$$

$$x = \frac{5}{12}$$

$$\therefore \quad x = \frac{5}{12} \text{ and } y = \frac{1}{3}$$

13. Solve simultaneously, the equations: $4^{x-2y} = 64$ and $25^{4x-3y} = 625$

Solution

$$4^{x-2y} = 64$$

Expressing both sides of the equation in the same base gives:

$$4^{x-2y} = 4^3$$

Equating the powers gives:

$$x - 2y = 3 \text{ ....Equation (1)}$$

Similarly, $25^{4x-3y} = 625$ can be expressed in base 25 as follows:

$$25^{4x-3y} = 25^2$$

Equating the powers gives:

$$4x + 3y = 2 \text{ ....Equation (2)}$$

From equation (1), $x = 3 + 2y$ ....Equation (3)

Substitute $3 + 2y$ for x in equation (2). This gives:

$$4x + 3y = 2 \text{ ....Equation (2)}$$

$$4(3 + 2y) + 3y = 2$$

$$12 + 8y + 3y = 2$$

$$11y = 2 - 12$$

$$11y = -10$$

$$y = \frac{-10}{11}$$

$$y = -\frac{10}{11}$$

Substitute $-\frac{10}{11}$ for y in equation (3) (Note that any of the equations can be used)

$$x = 3 + 2y \text{ ....Equation (3)}$$

$$= 3 + 2(-\frac{10}{11})$$

$$= 3 - \frac{20}{11}$$

$$x = \frac{33 - 20}{11}$$

$$x = \frac{13}{11}$$

$$x = 1\frac{2}{11}$$

$$\therefore \quad x = 1\frac{2}{11} \text{ and } y = -\frac{10}{11}$$

Exercise 6

1. Solve: $2^{5x+1} = 32^{2x-3}$

2. Solve the equation: $\dfrac{4^{2x+5}}{16^{2x-7}} = 64^{3x+2}$

3. Solve: $\dfrac{6^{3x-5}}{216^{x}} = \dfrac{216^{1-3x}}{36^{2x-3}}$

4. Solve the equation $5^{4x} = 15$

5. Solve the equation $3^{2x-11} = 20^{x+7}$

6. Solve $4.5^{2x} \times 10.2^{2-5x} = 8^{9x-4}$

7. Solve the equation $5^{x^2} = 2^{5x+2}$

8. Solve the equation $25^{x^2} = 125^{3x-1}$

9. Solve the equation: $4^x - 3 \times 2^x - 40 = 0$

10. Solve the equation $3^{2x} - 3^{x-1} - 78 = 0$

11. Solve: $5^{1+2x} + 2 \times 3^x - 7 = 0$

12. Solve the simultaneous equation: $9^{x-y} = 81$ and $25^{2x+y} = 25$

13. Solve simultaneously, the equations: $2^{5x-3y} = 32$ and $27^{2x-y} = 9^x$

14. Solve the simultaneous equation: $4^{x-2y} = 64$ and $25^{2x+5y} = 625$

15. Solve the equation $64^{x^2-2} = 16^{x+1}$

CHAPTER 7
ROOTS OF QUADRATIC EQUATIONS (USE OF ALPHA AND BETA)

A quadratic equation can have three cases of roots (solution of a quadratic equation). Consider a quadratic equation given by:

$$ax^2 + bx + c = 0$$

1. The roots are real and different if $b^2 - 4ac > 0$
2. The roots are real and equal if $b^2 - 4ac = 0$ or $b^2 = 4ac$
3. The roots are complex if $b^2 - 4ac < 0$

If each term of the quadratic equation above is divided by 'a', it gives:

$$x^2 + \frac{b}{a}x + \frac{c}{a} = 0 \ldots\ldots\ldots\ldots (1)$$

If α and β are the roots of the equation, then the equation can be written as:

$$x^2 - (\alpha + \beta)x + \alpha\beta = 0 \ldots\ldots\ldots\ldots (2)$$

Or

$$x^2 - (\text{sum of roots})x + (\text{product of roots}) = 0 \ldots\ldots\ldots\ldots(3)$$

Comparing equations (1) to (3) above shows that:

Sum of roots = $\alpha + \beta = -\dfrac{b}{a}$

And product of roots = $\alpha\beta = \dfrac{c}{a}$

The roots of quadratic equations can be expressed as functions of α and β.

In order to apply α and β in quadratic equations, it is important to know the following identities.

1. $\alpha^2 + \beta^2 = (\alpha + \beta)^2 - 2\alpha\beta$
2. $(\alpha - \beta)^2 = (\alpha + \beta)^2 - 4\alpha\beta$
3. $\alpha - \beta = \sqrt{(\alpha + \beta)^2 - 4\alpha\beta}$
4. $\alpha^2 - \beta^2 = (\alpha + \beta)(\alpha - \beta)$
   $$= (\alpha + \beta)\sqrt{(\alpha + \beta)^2 - 4\alpha\beta}$$
5. $\alpha^3 - \beta^3 = (\alpha - \beta)^3 + 3\alpha\beta(\alpha - \beta)$
6. $\alpha^3 + \beta^3 = (\alpha + \beta)^3 - 3\alpha\beta(\alpha + \beta)$

Examples

1. If α and β are the roots of the quadratic equation $2x^2 - 7x + 3 = 0$, find:
   a. $\alpha + \beta$
   b. $\alpha\beta$
   c. $\alpha^2 + \beta^2$
   d. $\dfrac{\alpha}{\beta} + \dfrac{\beta}{\alpha}$

e. $\dfrac{1}{\alpha} + \dfrac{1}{\beta}$

f. $\dfrac{1}{\alpha^2} + \dfrac{1}{\beta^2}$

g. $\alpha^3 + \beta^3$

Solution

a. $2x^2 - 7x + 3 = 0$

Recall the form: $ax^2 + bx + c = 0$

Comparing the two equations above shows that:

$\qquad$ a = 2, b = –7, c = 3

But $\alpha + \beta = -\dfrac{b}{a} = -(\dfrac{-7}{2})$ $\qquad$ (Take note of the use of negative sign)

$\therefore \quad \alpha + \beta = \dfrac{7}{2}$

And $\alpha\beta = \dfrac{c}{a} = \dfrac{3}{2}$

b. $\alpha\beta = \dfrac{3}{2}$ as shown in (a) above

c. $\alpha^2 + \beta^2$

Recall that $\alpha^2 + \beta^2 = (\alpha + \beta)^2 - 2\alpha\beta$, as given in the identities above.

Hence, $\alpha^2 + \beta^2 = (\alpha + \beta)^2 - 2\alpha\beta$

$\qquad\qquad\qquad = (\dfrac{7}{2})^2 - 2(\dfrac{3}{2})$ $\quad$ (By substituting $\dfrac{7}{2}$ for $(\alpha + \beta)$ and $\dfrac{3}{2}$ for $\alpha\beta$)

$\qquad\qquad\qquad = \dfrac{49}{4} - 3$

$\qquad\qquad\qquad = \dfrac{49 - 12}{4}$

$\qquad\qquad\qquad = \dfrac{37}{4}$

d. $\dfrac{\alpha}{\beta} + \dfrac{\beta}{\alpha} = \dfrac{\alpha^2 + \beta^2}{\alpha\beta}$ $\qquad$ (Note that the LCM of α and β is $\alpha\beta$)

$\qquad\qquad = \dfrac{(\alpha + \beta)^2 - 2\alpha\beta}{\alpha\beta}$ $\qquad$ (Note that $\alpha^2 + \beta^2 = (\alpha + \beta)^2 - 2\alpha\beta$)

$\qquad\qquad = \dfrac{\dfrac{37}{4}}{\dfrac{3}{2}}$ $\qquad$ (Note that from (c) above $\alpha^2 + \beta^2 = \dfrac{37}{4}$)

$\qquad\qquad = \dfrac{37}{4} \times \dfrac{2}{3}$

$$= \frac{37}{2} \times \frac{1}{3} \qquad \text{(After division by 2)}$$

$$= \frac{37}{6}$$

e. $\dfrac{1}{\alpha} + \dfrac{1}{\beta} = \dfrac{\beta + \alpha}{\alpha\beta}$

$$= \frac{\alpha + \beta}{\alpha\beta}$$

$$= \frac{\frac{7}{2}}{\frac{3}{2}}$$

$$= \frac{7}{2} \times \frac{2}{3}$$

$$= \frac{7}{3} \qquad \text{(After 2 cancels out)}$$

f. $\dfrac{1}{\alpha^2} + \dfrac{1}{\beta^2} = \dfrac{\beta^2 + \alpha^2}{\alpha^2\beta^2}$

$$= \frac{\alpha^2 + \beta^2}{(\alpha\beta)^2}$$

$$= \frac{\frac{37}{4}}{(\frac{3}{2})^2}$$

$$= \frac{\frac{37}{4}}{\frac{9}{4}}$$

$$= \frac{37}{4} \times \frac{4}{9}$$

$$= \frac{37}{9} \qquad \text{(Since 4 cancels out)}$$

g. $\alpha^3 + \beta^3 = (\alpha + \beta)^3 - 3\alpha\beta(\alpha + \beta)$

$$= (\frac{7}{2})^3 - 3(\frac{3}{2})(\frac{7}{2})$$

$$= \frac{343}{8} - \frac{9}{2}(\frac{7}{2})$$

$$= \frac{343}{8} - \frac{63}{4}$$

$$= \frac{343 - 126}{8}$$

$$= \frac{217}{8}$$

2. If the roots of the quadratic equation, $3x^2 + 5x - 9 = 0$, are α and β, find the equation whose roots are α^2 and β^2

<u>Solution</u>

$$3x^2 + 5x - 9 = 0$$

From this equation: a = 3, b = 5 and c = −9

But, $\alpha + \beta = -\dfrac{b}{a}$

$\qquad = -\dfrac{5}{3}$ (Since a = 3 and b = 5)

Also, $\alpha\beta = \dfrac{c}{a}$

$\qquad = \dfrac{-9}{3}$ (Since a = 3 and c = −9)

$\qquad = -3$

Now, roots of new equation are α^2 and β^2

∴   Sum of roots of new equation $= \alpha^2 + \beta^2$

$\qquad = (\alpha + \beta)^2 - 2\alpha\beta$ (From the identities given above)

$\qquad = (-\dfrac{5}{3})^2 - 2(-3)$

$\qquad = \dfrac{25}{9} + 6$

$\qquad = \dfrac{25 + 54}{9}$

Sum of roots $(\alpha + \beta) = \dfrac{79}{9}$

And product of roots of new equation $= \alpha^2 \times \beta^2$

$\qquad = \alpha^2\beta^2$

$\qquad = (\alpha\beta)^2$

$\qquad = (-3)^2$

Product of roots $(\alpha\beta) = 9$

Recall that a quadratic equation is represented as:

$\quad x^2 - (\text{sum of roots})x + (\text{product of roots}) = 0$

Or, $x^2 - (\alpha + \beta)x + (\alpha\beta) = 0$

When the values obtained for $\alpha + \beta$ and $\alpha\beta$ are substituted into the equation above, it gives:

$$x^2 - \dfrac{79}{9}x + 9 = 0$$

When each term is multiplied by 9 in order to clear out the fraction, it gives:

$$9x^2 - 79x + 81 = 0$$

∴ The equation whose roots are α^2 and β^2 is $9x^2 - 79x + 81 = 0$

3. If the roots of the quadratic equation $5x^2 + x + 4 = 0$, are α and β, find the equation whose roots are $\dfrac{\alpha}{\beta}$ and $\dfrac{\beta}{\alpha}$

Solution

$$5x^2 + x + 4 = 0$$

From this equation: $a = 5$, $b = 1$ and $c = 4$

But, $\alpha + \beta = -\dfrac{b}{a}$

$\qquad\qquad = -\dfrac{1}{5}$ (Since $a = 5$ and $b = 1$)

Also, $\alpha\beta = \dfrac{c}{a}$

$\qquad\quad = \dfrac{4}{5}$ (Since $a = 5$ and $c = 4$)

Now, roots of new equation are $\dfrac{\alpha}{\beta}$ and $\dfrac{\beta}{\alpha}$

$\therefore$ Sum of roots of new equation $= \dfrac{\alpha}{\beta} + \dfrac{\beta}{\alpha}$

$$= \frac{\alpha^2 + \beta^2}{\alpha\beta} \qquad \text{(From the identities given above)}$$

$$= \frac{(\alpha + \beta)^2 - 2\alpha\beta}{\alpha\beta}$$

$$= \frac{\left(-\frac{1}{5}\right)^2 - 2\left(\frac{4}{5}\right)}{\frac{4}{5}}$$

$$= \frac{\frac{1}{25} - \frac{8}{5}}{\frac{4}{5}}$$

$$= \frac{\frac{1-40}{25}}{\frac{4}{5}}$$

$$= \frac{-\frac{39}{25}}{\frac{4}{5}}$$

$$= -\frac{39}{25} \times \frac{5}{4}$$

$$= -\frac{39}{5} \times \frac{1}{4} \qquad \text{(After division by 5)}$$

$$= -\frac{39}{20}$$

Sum of roots $(\alpha + \beta) = -\dfrac{39}{20}$

And product of roots of new equation $= \dfrac{\alpha}{\beta} \times \dfrac{\beta}{\alpha}$

$$= \frac{\alpha\beta}{\alpha\beta}$$

$$= 1$$

Product of roots $(\alpha\beta) = 1$

Recall that a quadratic equation is represented as:

$x^2 - $ (sum of roots)$x + $ (product of roots) $= 0$

Or, $x^2 - (\alpha + \beta)x + (\alpha\beta) = 0$

When the values obtained for $\alpha + \beta$ and $\alpha\beta$ are substituted into the equation above, it gives:

$$x^2 - (-\frac{39}{20})x + 1 = 0$$

$$x^2 + \frac{39}{20}x + 1 = 0$$

When each term is multiplied by 20 in order to clear out the fraction, it gives:

$$20x^2 + 39x + 20 = 0$$

$\therefore$ The equation whose roots are $\dfrac{\alpha}{\beta}$ and $\dfrac{\beta}{\alpha}$ is $20x^2 + 39x + 20 = 0$

4. Given that the quadratic equation $2x^2 - 5x - 3 = 0$, has roots α and β, find the equation whose roots are $\dfrac{1}{\alpha}$ and $\dfrac{1}{\beta}$

Solution

$$2x^2 - 5x - 3 = 0$$

From this equation: a = 2, b = –5 and c = –3

But, $\alpha + \beta = -\dfrac{b}{a}$

$\qquad = -(\dfrac{-5}{2})$ (Since a = 2 and b = –5)

$\qquad = \dfrac{5}{2}$

Also, $\alpha\beta = \dfrac{c}{a}$

$\qquad = \dfrac{-3}{2}$ (Since a = 2 and c = –3)

Now, roots of new equation are $\dfrac{1}{\alpha}$ and $\dfrac{1}{\beta}$

$\therefore$ Sum of roots of new equation $= \dfrac{1}{\alpha} + \dfrac{1}{\beta}$

$\qquad = \dfrac{\beta + \alpha}{\alpha\beta}$

$\qquad = \dfrac{\alpha + \beta}{\alpha\beta}$

$\qquad = \dfrac{\dfrac{5}{2}}{-\dfrac{3}{2}}$

$\qquad = \dfrac{5}{2} \times (-\dfrac{2}{3})$

71

$$= -\frac{5}{3}$$

Sum of roots of new equation $(\alpha + \beta) = -\frac{5}{3}$

And product of roots of new equation $= \frac{1}{\alpha} \times \frac{1}{\beta}$

$$= \frac{1}{\alpha\beta}$$

$$= \frac{1}{-\frac{3}{2}}$$

$$= \frac{1}{1} \times (-\frac{2}{3})$$

Product of roots of new equation $(\alpha\beta) = -\frac{2}{3}$

Recall that a quadratic equation is represented as:

$$x^2 - (\text{sum of roots})x + (\text{product of roots}) = 0$$

Or, $x^2 - (\alpha + \beta)x + (\alpha\beta) = 0$

When the values obtained for $\alpha + \beta$ and $\alpha\beta$ are substituted into the equation above, it gives:

$$x^2 - (-\frac{5}{3})x + (-\frac{2}{3}) = 0$$

$$x^2 + \frac{5}{3}x - \frac{2}{3} = 0$$

When each term is multiplied by 3 in order to clear out the fraction, it gives:

$$3x^2 + 5x - 2 = 0$$

∴ The equation whose roots are $\frac{1}{\alpha}$ and $\frac{1}{\beta}$ is $3x^2 + 5x - 2 = 0$

5. If α and β are the roots of the quadratic equation $x^2 - 7x + 9 = 0$, find the quadratic equation whose roots are $\frac{1}{\alpha^2}$ and $\frac{1}{\beta^2}$

<u>Solution</u>

$$x^2 - 7x + 9 = 0$$

From this equation: a = 1, b = −7 and c = 9

But, $\alpha + \beta = -\frac{b}{a}$

$$= -(\frac{-7}{1}) \qquad (\text{Since a = 1 and b = −7})$$

$$= 7$$

Also, $\alpha\beta = \frac{c}{a}$

$$= \frac{9}{1} \qquad (\text{Since a = 1 and c = 9})$$

$$= 9$$

Now, roots of new equation are $\dfrac{1}{\alpha^2}$ and $\dfrac{1}{\beta^2}$

∴ Sum of roots of new equation $= \dfrac{1}{\alpha^2} + \dfrac{1}{\beta^2}$

$= \dfrac{\alpha^2 + \beta^2}{\alpha^2 \beta^2}$

$= \dfrac{(\alpha + \beta)^2 - 2\alpha\beta}{(\alpha\beta)^2}$

$= \dfrac{7^2 - 2(9)}{9^2}$

$= \dfrac{49 - 18}{81}$

Sum of roots $(\alpha + \beta) = \dfrac{31}{81}$

And product of roots of new equation $= \dfrac{1}{\alpha^2} \times \dfrac{1}{\beta^2}$

$= \dfrac{1}{\alpha^2 \beta^2}$

$= \dfrac{1}{(\alpha\beta)^2}$

$= \dfrac{1}{9^2}$

Product of roots $(\alpha\beta) = \dfrac{1}{81}$

Recall that a quadratic equation is represented as:

$x^2 - (\text{sum of roots})x + (\text{product of roots}) = 0$

Or, $x^2 - (\alpha + \beta)x + (\alpha\beta) = 0$

When the values obtained for $\alpha + \beta$ and $\alpha\beta$ are substituted into the equation above, it gives:

$x^2 - \left(\dfrac{31}{81}\right)x + \dfrac{1}{81} = 0$

When each term is multiplied by 81 in order to clear out the fractions, it gives:

$81x^2 - 31x + 1 = 0$

∴ The equation whose roots are $\dfrac{1}{\alpha^2} + \dfrac{1}{\beta^2}$ is $81x^2 - 31x + 1 = 0$

6. If α and β are the roots of the quadratic equation $2x^2 + 3x + 5 = 0$, find the quadratic equation whose roots are α^3 and β^3

Solution

$2x^2 + 3x + 5 = 0$

From this equation: a = 2, b = 3 and c = 5

But, $\alpha + \beta = -\dfrac{b}{a}$

$\quad = -\dfrac{3}{2}$

Also, $\alpha\beta = \dfrac{c}{a}$

$\quad = \dfrac{5}{2}$

Now, roots of new equation are α^3 and β^3

$\therefore \quad$ Sum of roots of new equation $= \alpha^3 + \beta^3$

$\qquad = (\alpha + \beta)^3 - 3\alpha\beta(\alpha + \beta)$

$\qquad = (-\dfrac{3}{2})^3 - 3(\dfrac{5}{2})(-\dfrac{3}{2})$

$\qquad = -\dfrac{27}{8} + \dfrac{45}{4}$

$\qquad = \dfrac{-27 + 90}{8}$

$\qquad = \dfrac{63}{8}$

Sum of roots $(\alpha + \beta) = \dfrac{63}{8}$

And product of roots of new equation $= \alpha^3 \times \beta^3$

$\qquad = (\alpha\beta)^3$

$\qquad = (\dfrac{5}{2})^3$

$\qquad = \dfrac{125}{8}$

Product of roots of new equation $(\alpha\beta) = \dfrac{125}{8}$

Recall that a quadratic equation is represented as:

$\quad x^2 - (\text{sum of roots})x + (\text{product of roots}) = 0$

Or, $x^2 - (\alpha + \beta)x + (\alpha\beta) = 0$

When the values obtained for $\alpha + \beta$ and $\alpha\beta$ are substituted into the equation above, it gives:

$\qquad x^2 - (\dfrac{63}{8})x + \dfrac{125}{8} = 0$

When each term is multiplied by 8 in order to clear out the fractions, it gives:

$\qquad 8x^2 - 63x + 125 = 0$

$\therefore$ The equation whose roots are α^3 and β^3 is $8x^2 - 63x + 125 = 0$

7. If α and β are the roots of the equation $4x^2 - 12x + 7 = 0$, find the values:

a. $\alpha - \beta$

b. $\alpha^2 - \beta^2$

Solution

$$4x^2 - 12x + 7 = 0$$

From this equation: a = 4, b = –12 and c = 7

But, $\alpha + \beta = -\dfrac{b}{a}$

$$= -\left(\dfrac{-12}{4}\right)$$

$$= 3$$

Also, $\alpha\beta = \dfrac{c}{a}$

$$= \dfrac{7}{4}$$

Recall that: $(\alpha - \beta)^2 = (\alpha + \beta)^2 - 4\alpha\beta$

Taking the square root of both sides gives $\alpha - \beta$ as follows:

$$\alpha - \beta = \sqrt{(\alpha + \beta)^2 - 4\alpha\beta}$$

$$= \sqrt{3^2 - 4\left(\dfrac{7}{4}\right)}$$

$$= \sqrt{9 - 7}$$

$$= \sqrt{2}$$

$\therefore \ \alpha - \beta = \sqrt{2}$ or 1.41

b. Recall that: . $\alpha^2 - \beta^2 = (\alpha + \beta)(\alpha - \beta)$ (This is a difference of two squares)

$\qquad = (3)(\sqrt{2})$ (Since $\alpha + \beta = 3$ and $\alpha - \beta = \sqrt{2}$ from (a) above)

$\qquad = 3\sqrt{2}$

$\therefore \ \alpha^2 - \beta^2 = 3\sqrt{2}$ or 4.24

8. If α and β are the roots of the equation $3x^2 + 11x - 8 = 0$, find the value of $(\alpha - \beta)^2$

Solution

$$3x^2 + 11x - 8 = 0$$

From this equation: a = 3, b = 11 and c = –8

But, $\alpha + \beta = -\dfrac{b}{a}$

$$= -\dfrac{11}{3}$$

Also, $\alpha\beta = \dfrac{c}{a}$

$$= -\dfrac{8}{3}$$

Recall that: $(\alpha - \beta)^2 = (\alpha + \beta)^2 - 4\alpha\beta$

$$= (-\frac{11}{3})^2 - 4(-\frac{8}{3})$$

$$= \frac{121}{9} + \frac{32}{3}$$

$$= \frac{121 + 96}{9}$$

$$= \frac{217}{9}$$

$$\therefore \quad (\alpha - \beta)^2 = \frac{217}{9}$$

9. If the roots of the equation $2x^2 - 3x - 1 = 0$ are α and β, find the value of $(\alpha - \beta)^3$

<u>Solution</u>

$$2x^2 - 3x - 1 = 0$$

From this equation: a = 2, b = –3 and c = –1

But, $\alpha + \beta = -\dfrac{b}{a}$

$$= -(\frac{-3}{2})$$

$$= \frac{3}{2}$$

Also, $\alpha\beta = \dfrac{c}{a}$

$$= -\frac{1}{2}$$

Let us first find the value of $\alpha - \beta$ in order to determine $(\alpha - \beta)^3$

Recall that: $\alpha - \beta = \sqrt{(\alpha + \beta)^2 - 4\alpha\beta}$

$$= \sqrt{(\frac{3}{2})^2 - 4(-\frac{1}{2})}$$

$$= \sqrt{\frac{9}{4} + 2}$$

$$= \sqrt{\frac{9 + 8}{4}}$$

$$= \sqrt{\frac{17}{4}}$$

$$\alpha - \beta = 2.0616$$

$$\therefore \quad (\alpha - \beta)^3 = (2.0616)^3$$

$$= 8.76$$

10. If α and β are the roots of the quadratic equation $3x^2 - 5x - 1 = 0$, form an equation whose roots are:

a. $(\alpha^2 + \frac{1}{\beta})$ and $(\beta^2 + \frac{1}{\alpha})$

b. $(\alpha - \frac{1}{\beta})$ and $(\beta - \frac{1}{\alpha})$

<u>Solution</u>

$\qquad 3x^2 - 5x - 1 = 0$

From this equation: $a = 3$, $b = -5$ and $c = -1$

Hence, $\alpha + \beta = -\dfrac{b}{a}$

$\qquad = -(\dfrac{-5}{3})$

$\qquad = \dfrac{5}{3}$

Also, $\alpha\beta = \dfrac{c}{a}$

$\qquad = -\dfrac{1}{3}$

Now, roots of new equation are $(\alpha^2 + \frac{1}{\beta})$ and $(\beta^2 + \frac{1}{\alpha})$

∴ Sum of roots of new equation $= \alpha^2 + \dfrac{1}{\beta} + \beta^2 + \dfrac{1}{\alpha}$

$\qquad = \alpha^2 + \beta^2 + \dfrac{1}{\beta} + \dfrac{1}{\alpha}$ (By rearrangement)

$\qquad = (\alpha + \beta)^2 - 2\alpha\beta + \dfrac{\alpha + \beta}{\alpha\beta}$ (Note that $\alpha^2 + \beta^2 = (\alpha + \beta)^2 - 2\alpha\beta$)

$\qquad = (\dfrac{5}{3})^2 - 2(-\dfrac{1}{3}) + \dfrac{\frac{5}{3}}{-\frac{1}{3}}$

$\qquad = \dfrac{25}{9} + \dfrac{2}{3} - (\dfrac{5}{3} \times \dfrac{3}{1})$

$\qquad = \dfrac{25}{9} + \dfrac{2}{3} - 5$

$\qquad = \dfrac{25 + 6 - 45}{9}$

$\qquad = -\dfrac{14}{9}$

Sum of roots $(\alpha + \beta) = -\dfrac{14}{9}$

And product of roots of new equation $= (\alpha^2 + \frac{1}{\beta})(\beta^2 + \frac{1}{\alpha})$

$\qquad = \alpha^2\beta^2 + \dfrac{\alpha^2}{\alpha} + \dfrac{\beta^2}{\beta} + \dfrac{1}{\alpha\beta}$

$$= (\alpha\beta)^2 + \alpha + \beta + \frac{1}{\alpha\beta}$$

$$= (-\frac{1}{3})^2 + \frac{5}{3} + \frac{1}{-\frac{1}{3}}$$

$$= \frac{1}{9} + \frac{5}{3} - 3$$

$$= \frac{1 + 15 - 27}{9}$$

$$= -\frac{11}{9}$$

Product of roots of new equation $(\alpha\beta) = -\frac{11}{9}$

Recall that a quadratic equation is represented as:

$$x^2 - (\text{sum of roots})x + (\text{product of roots}) = 0$$

Or, $x^2 - (\alpha + \beta)x + (\alpha\beta) = 0$

When the values obtained for $\alpha + \beta$ and $\alpha\beta$ are substituted into the equation above, it gives:

$$x^2 - (-\frac{14}{9})x - \frac{11}{9} = 0$$

When each term is multiplied by 9 in order to clear out the fractions, it gives:

$$9x^2 + 14x - 11 = 0$$

∴ The equation whose roots are $(\alpha^2 + \frac{1}{\beta})$ and $(\beta^2 + \frac{1}{\alpha})$ is $9x^2 + 14x - 11 = 0$

b. Since roots of new equation are $(\alpha - \frac{1}{\beta})$ and $(\beta - \frac{1}{\alpha})$,

∴ Sum of roots of new equation $= \alpha - \frac{1}{\beta} + \beta - \frac{1}{\alpha}$

$$= \alpha + \beta - (\frac{1}{\beta} + \frac{1}{\alpha}) \quad \text{(By rearrangement)}$$

$$= \alpha + \beta - (\frac{\alpha + \beta}{\alpha\beta})$$

$$= \frac{5}{3} - \frac{\frac{5}{3}}{-\frac{1}{3}}$$

$$= \frac{5}{3} + (\frac{5}{3} \times \frac{3}{1})$$

$$= \frac{5}{3} + 5$$

$$= \frac{5 + 15}{3}$$

Sum of roots $(\alpha + \beta) = \frac{20}{3}$

And product of roots of new equation $= (\alpha - \frac{1}{\beta})(\beta - \frac{1}{\alpha})$

$$= \alpha\beta - \frac{\alpha}{\alpha} - \frac{\beta}{\beta} + \frac{1}{\alpha\beta}$$

$$= \alpha\beta - 1 - 1 + \frac{1}{\alpha\beta}$$

$$= \alpha\beta - 2 + \frac{1}{\alpha\beta}$$

$$= -\frac{1}{3} - 2 + \frac{1}{-\frac{1}{3}}$$

$$= -\frac{1}{3} - 2 - 3$$

$$= -\frac{1}{3} - 5$$

$$= \frac{-1 - 15}{3}$$

$$= -\frac{16}{3}$$

Product of roots of new equation $(\alpha\beta) = -\frac{16}{3}$

Recall that a quadratic equation is represented as:

$x^2 -$ (sum of roots)$x +$ (product of roots) $= 0$

Or, $x^2 - (\alpha + \beta)x + (\alpha\beta) = 0$

When the values obtained for $\alpha + \beta$ and $\alpha\beta$ are substituted into the equation above, it gives:

$$x^2 - (\frac{20}{3})x - \frac{16}{3} = 0$$

When each term is multiplied by 3 in order to clear out the fractions, it gives:

$$3x^2 - 20x - 16 = 0$$

$\therefore$ The equation whose roots are $(\alpha - \frac{1}{\beta})$ and $(\beta - \frac{1}{\alpha})$ is $3x^2 - 20x - 16 = 0$

Maximum and Minimum Values of a Quadratic Function

The maximum or minimum value of a quadratic function is given by:

$$\frac{4ac - b^2}{4a}$$

The equation of the line of symmetry of a quadratic curve is given by: $x = -\frac{b}{2a}$

This line is called the axis of symmetry of a quadratic curve.

For a quadratic equation to have equal roots, the following condition must be met:

$b^2 = 4ac$ (From $b^2 - 4ac = 0$)

A quadratic equation which has equal roots is also a perfect square. Therefore, if a quadratic equation is a perfect square, then it follows that:

$b^2 = 4ac$

Examples

1. If $(2x + p)(3x - q) = 6x^2 - 7x + 2$, where p and q are constant, find the possible values of p.

<u>Solution</u>

$$(2x + p)(3x - q) = 6x^2 - 7x + 2$$

Let us expand the left hand side. This gives

$6x^2 - 2qx + 3px - pq = 6x^2 - 7x + 2$

$6x^2 - (2q - 3p)x - pq = 6x^2 - 7x + 2$ [Note: $-2qx + 3px$ has been factorized to give $-(2q - 3p)x$]

Comparing the terms in x on both sides of the equation shows that:

$-(2p - 3p) = -7$ (The terms in x)

$\therefore$ $2p - 3p = 7$ (After dividing both sides by -1)

$\therefore$ $2p - 3p = 7$ ....Equation (1)

Similarly, comparing the constant terms on both sides of the equation shows that:

$-pq = 2$

$\therefore$ $pq = -2$ (After dividing both sides by -1)

$pq = -2$ ....Equation(2)

From equation (2):

$$q = -\frac{2}{p}$$

Substitute $-\frac{2}{p}$ for q in equation (1)

$2q - 3p = 7$ ....Equation(1)

$2(-\frac{2}{p}) - 3p = 7$

$-\frac{4}{p} - 3p = 7$

Multiply each term by p in order to clear the fraction.

$p(-\frac{4}{p}) - p(3p) = p(7)$

$-4 - 3p^2 = 7p$

$0 = 3p^2 + 7p + 4$

$\therefore$ $3p^2 + 7p + 4 = 0$

Solving this equation by factorization method gives:

$3p^2 + 3p + 4p + 4 = 0$

$3p(p + 1) + 4(p + 1) = 0$

$\therefore$ $(p + 1)(3p + 4) = 0$

Equating each bracket to zero and solving each equation gives $p = -1$ or $p = -\frac{4}{3}$

$\therefore$ The possible values of p are -1 and $-\frac{4}{3}$

2. If $2x^2 - (m - 4)x - 4(m + 2) = 0$ has equal roots, find the possible values of m

Solution

$2x^2 - (m - 4)x - 4(m + 2) = 0$

From the equation, the values of a, b and c are given by a = 2, b = $-(m - 4)$, c = $-4(m + 2)$

Or, a = 2, b = $-m + 4$, c = $-4m - 8$ (When we expand the brackets above)

For a quadratic equation to have equal roots, the following condition must be met:

$b^2 = 4ac$ (From $b^2 - 4ac = 0$)

Substituting the values above gives:

$b^2 = 4ac$

$(-m + 4)^2 = 4 \times 2 \times (-4m - 8)$ (By substituting the values of a, b and c)

$(4 - m)^2 = 8(-4m - 8)$ (Note that $-m + 4 = 4 - m$)

$(4 - m)(4 - m) = -32m - 64$

$16 - 4m - 4m + m^2 = -32m - 64$

$m^2 - 8m + 32m + 16 + 64 = 0$

$m^2 + 24m + 80 = 0$

∴ $(m + 20)(m + 4) = 0$ (After factorization)

∴   m = -20 or m = -4

Hence, the possible values of m are -20 and -4.

3. Find the value of the constant k for which the expression $4x^2 + 20x + (12 + k)$ is a perfect square.

Solution

$4x^2 + 20x + (12 + k)$

From this equation, a, b and c are:

a = 4, b = 20, c = 12 + k

For a quadratic equation to be a perfect square, the condition below must be met.

$b^2 = 4ac$

$20^2 = 4 \times 4 \times (12 + k)$

$400 = 16(12 + k)$

$400 = 192 + 16K$

$400 - 192 = 16K$

∴ $k = \dfrac{208}{16}$

k = 13

4. Find the maximum value of $5 - 11x - 2x^2$

<u>Solution</u>

Recall that maximum value is given by:

$$\frac{4ac - b^2}{4a}$$

From the equation $5 - 11x - 2x^2$:

$a = -2, b = -11, c = 5$

$\therefore$ Maximum value $= \dfrac{4ac - b^2}{4a}$

$$= \frac{4(-2 \times 5) - (-11)^2}{4(-2)}$$

$$= \frac{-40 - 121}{-8}$$

$$= \frac{-161}{-8}$$

$$= \frac{161}{8}$$

5. Given the function, $y = 3x^2 + 7x - 10$, find:

a. the minimum value of y

b. the value of x for which y is minimum

<u>Solution</u>

a. $y = 3x^2 + 7x - 10$

From the equation:

$a = 3, b = 7, c = -10$

The minimum value of y is given by:

$$y = \frac{4ac - b^2}{4a}$$

$$= \frac{4(3)(-10) - (7)^2}{4(3)}$$

$$= \frac{-120 - 49}{12}$$

$$= \frac{-169}{12}$$

$$= -\frac{169}{12}$$

b. The value of x for which y is minimum is given by:

$$x = \frac{-b}{2a}$$

$$= \frac{-7}{2(3)}$$

$x = \dfrac{-7}{6}$ (Note that this is also the line of symmetry of the quadratic curve)

6. Given the function, $y = 12 - 9x - 2x^2$, find:
a. the maximum value of y
b. the value of x for which y is maximum

Solution

a. $y = 12 - 9x - 2x^2$

From the equation:

  $a = -2, b = -9, c = 12$

The maximum value of y is given by:

$$y = \frac{4ac - b^2}{4a}$$

$$= \frac{4(-2)(12) - (-9)^2}{4(-2)}$$

$$= \frac{-96 - 81}{-8}$$

$$= \frac{-177}{-8}$$

$$= \frac{177}{8}$$

b. The value of x for which y is maximum is given by:

$$x = \frac{-b}{2a}$$

$$= \frac{-(-9)}{2(-2)}$$

$$x = \frac{9}{-4}$$ (Note that this is also the line of symmetry of the quadratic curve)

$$x = -\frac{9}{4}$$

7. One root of the quadratic equation $x^2 - (4 + p)x + 12 = 0$ is three times the other. Find:
a. the roots of the equation
b. the possible values of p

Solution

a. $x^2 - (4 + p)x + 12 = 0$

Let one of the roots be α. Therefore, the other root is 3α.

 From the equation:

Product of roots = 12   (From $\dfrac{c}{a}$ in the equation. Note that a = 1)

$\therefore$ $\alpha(3\alpha) = 12$

$3\alpha^2 = 12$

$\alpha^2 = 4$ (After dividing both sides by 3)

$\therefore$ $\alpha = \sqrt{4}$

$\alpha = \pm 2$

$\therefore$ $\alpha = 2$ or -2

When $\alpha = 2$, the other root is $3\alpha = 3 \times 2 = 6$

When $\alpha = -2$, the other root is $3\alpha = 3(-2) = -6$

Therefore, the roots are either 2 and 6, or -6 and -2

b. From the equation: $x^2 - (4 + p)x + 12 = 0$:

     $a = 1, b = -(4 + p), c = 12$

Sum of the roots $= -b/a$

     $\alpha + 3\alpha = -[-(4 + p)]/1$ (Note that α and 3α are the two roots)

        $4\alpha = 4 + p$

     $4\alpha - 4 = p$

$\therefore$ When $\alpha = 2$, p is given by

      $4\alpha - 4 = p$

      $4(2) - 4 = p$

      $8 - 4 = p$

$\therefore$ $p = 4$

Also, when $\alpha = -2$, p is given by:

      $4\alpha - 4 = p$

      $4(-2) - 4 = p$

      $-8 - 4 = p$

      $p = -12$

$\therefore$ The possible values of p are 4 and -12.

Exercise 7

1. If α and β are the roots of the quadratic equation $3x^2 - 8x + 5 = 0$, find:

a. $\alpha + \beta$

b. $\alpha\beta$

c. $\alpha^2 + \beta^2$

d. $\dfrac{\alpha}{\beta} + \dfrac{\beta}{\alpha}$

e. $\dfrac{1}{\alpha} + \dfrac{1}{\beta}$

f. $\dfrac{1}{\alpha^2} + \dfrac{1}{\beta^2}$

g. $\alpha^3 + \beta^3$

2. If the roots of the quadratic equation, $2x^2 + 12x - 11 = 0$, are α and β, find the equation whose roots are α^2 and β^2

3. If the roots of the quadratic equation $4x^2 + 3x - 10 = 0$, are α and β, find the equation whose roots are $\dfrac{\alpha}{\beta}$ and $\dfrac{\beta}{\alpha}$

4. Given that the quadratic equation $x^2 - 15x - 2 = 0$, has roots α and β, find the equation whose roots are $\dfrac{1}{\alpha}$ and $\dfrac{1}{\beta}$

5. If α and β are the roots of the quadratic equation $3x^2 - 15x + 7 = 0$, find the quadratic equation whose roots are $\dfrac{1}{\alpha^2}$ and $\dfrac{1}{\beta^2}$

6. If α and β are the roots of the quadratic equation $2x^2 + 5x - 6 = 0$, find the quadratic equation whose roots are α^3 and β^3

7. If α and β are the roots of the equation $5x^2 - 8x + 2 = 0$, find the values of:

a. $\alpha - \beta$

b. $\alpha^2 - \beta^2$

8. If α and β are the roots of the equation $10x^2 + 15x - 8 = 0$, find the value of $(\alpha - \beta)^2$

9. If the roots of the equation $4x^2 - 2x - 3 = 0$ are α and β, find the value of $(\alpha - \beta)^3$

10. If α and β are the roots of the quadratic equation $x^2 - 2x - 3 = 0$, form an equation whose roots are:

a. $(\alpha^2 + \dfrac{1}{\beta})$ and $(\beta^2 + \dfrac{1}{\alpha})$

b. $(\alpha - \dfrac{1}{\beta})$ and $(\beta - \dfrac{1}{\alpha})$

11. If $(3x + m)(x - n) = 3x^2 - 11x + 6$, where m and n are constants, find the possible values of m and n.

12. If $2x^2 - (p + 3)x - 2(p - 2) = 0$ has equal roots, find the possible values of p.

13. Find the value of the constant k for which the expression $3x^2 + 4x - (8 + k)$ is a perfect square.

14. Find the maximum value of $7 - 14x - 5x^2$

15. Given the function, $y = 2x^2 + 11x - 12$, find:

a. the minimum value of y

b. the value of x for which y is minimum

16. Given the function, $y = 18 - 15x - 4x^2$, find:

a. the maximum value of y

b. the value of x for which y is maximum

17. One root of the quadratic equation $2x^2 - (2 - k)x + 6 = 0$ is three times the other. Find:

a. the roots of the equation

b. the possible values of k

18. If α and β are the roots of the equation $2x^2 - 6x + 3 = 0$, find the values of $\alpha^2 - \beta^2$

19. Find the value of the constant k for which the expression $5x^2 + 10x - (3 + k)$ is a perfect square.

20. One root of the quadratic equation $4x^2 - (9 - k)x + 10 = 0$ is twice the other. Find the roots of the equation.

CHAPTER 8
FUNCTIONS

A function, f, whose output is y, and input or variable is x, is a rule which describes how a value of x is used to obtain a value of y. A function is usually represented in the form of an equation as follows:

$$y = f(x)$$

The rule of a function must imply that for each input of x, there must be only one output of y. For example, the equation:

$$y = 5x - 1$$

shows that for every value of x, there is only one value of y. Hence this rule is a function. However, an equation such as:

$$y = \sqrt{x} \quad \text{or} \quad y = x^{\frac{1}{2}}$$

is a rule that can give two possible values of y. If $x = 4$, then $y = 4^{\frac{1}{2}}$. Which gives:

$$y = \sqrt{4}$$
$$y = \pm 2 \text{ which means } +2 \text{ or } -2.$$

This shows that for a positive value of x, there are two possible values of y. Hence, $y = x^{\frac{1}{2}}$ is not a function. A function should give only one value of the output y.

Examples

Determine if each of the following is a function or not.

1. $y = x^{\frac{1}{4}} - 2$
2. $y = 3x^2 - 5\sqrt[3]{x}$
3. $y = \sqrt{20 - x^2}$
4. $y = \dfrac{1}{2x - 1}$

<u>Solutions</u>

1. $y = x^{\frac{1}{4}} - 2$

In order to test if this is a function, we can put a value of x and see if we will get only one value of y. Let us take $x = 16$ and put it in the equation. This gives:

$$y = x^{\frac{1}{4}} - 2$$
$$y = 16^{\frac{1}{4}} - 2$$
$$= \sqrt[4]{16} - 2$$
$$= \pm 2 - 2$$

$= +2 - 2$ or $-2 - 2$

Hence, y = 0 or y = –4

Therefore, $y = x^{\frac{1}{4}} - 2$ is not a function since a value of x produces two values of y.

2. $y = 3x^2 - 5\sqrt[3]{x}$

Let us put $x = -8$ into the equation. Note that any convenient value of can be used.

$$y = 3x^2 - 5\sqrt[3]{x}$$
$$= 3(-8)^2 - 5\sqrt[3]{-8}$$
$$= 3(64) - 5(-2)$$
$$= 192 + 10$$
$$= 202$$

Hence, $y = 3x^2 - 5\sqrt[3]{x}$ is a function since only one value of y is produced.

3. $y = \sqrt{20 - x^2}$

Let us put $x = 3$ into the equation. Be careful to put in the square root sign, only values of x that will give a positive value. This is because we cannot evaluate values such as $\sqrt{-10}$, but only positive values such as $\sqrt{10}$

Hence, $y = \sqrt{20 - x^2}$
$$= \sqrt{20 - 3^2} \qquad \text{(Since } x = 3 \text{ as stated above)}$$
$$= \sqrt{20 - 9}$$
$$= \sqrt{11}$$
$$y = \pm\sqrt{11}$$

Therefore, $y = +\sqrt{11}$ or $-\sqrt{11}$

Hence, $y = \sqrt{20 - x^2}$ is not a function since there are two possible values is y.

4. $y = \dfrac{1}{2x - 1}$

Let us put $x = -2$ into the equation as follows:

$$y = \frac{1}{2x - 1}$$
$$= \frac{1}{2(-2) - 1}$$
$$= \frac{1}{-4 - 1}$$
$$= \frac{1}{-5}$$
$$y = -0.2$$

Hence, $y = \dfrac{1}{2x - 1}$ is a function

Domain and Range of a Function

Domain are all the values of x that a function can take and process.

Range is the set of values y obtained from each domain of a function. Range is also called co–domain.

For example, if y = $\sqrt{4 - x^2}$, where x and y are real numbers, and x is a whole number, then the domain is the set of values −2, −1, 0, 1, 2. These are the values of x that will not give negative values in the square root sign. The range (i.e. the corresponding values of y from the domain values of x) is the set of values 0, $\sqrt{3}$, 2, $\sqrt{3}$, 0, or simply 0, $\sqrt{3}$, 2, when avoiding repetition.

A function can be defined by a given domain of values.

For example if a function is defined as follows:

$$y = 2x, \quad -5 < x < 1$$

then from the defined values, the domain of the function is:

$$-5 < x < 1$$

When each of the extreme values (i.e. −5 and 1) is substituted into the function, then the range of the function will be obtained as:

$$y = 2(-5) = -10 \quad \text{(When } x = -5)$$

and $y = 2(1) = 2 \quad$ (When $x = 1$)

This gives the range: $\quad -10 < x < 2$

If a function is given by:

$$y = \frac{1}{(x-2)(x+5)}$$

then the domain can be any value of x except $x = 2$ and $x = -5$. These two values of x will give a denominator of zero which makes the function undefined. Let us see the result of these two values of x. When $x = 2$, then:

$$y = \frac{1}{(x-2)(x+5)}$$

$$= \frac{1}{(2-2)(2+5)}$$

$$= \frac{1}{(0)(7)} = \frac{1}{0} \quad \text{(This is undefined)}$$

When $x = -5$, then:

$$y = \frac{1}{(x-2)(x+5)}$$

$$= \frac{1}{(-5-2)(-5+5)}$$

$$= \frac{1}{(-7)(0)} = \frac{1}{0} \quad \text{(This is undefined)}$$

Hence, $x = 2$ and $x = -5$ are two values that should not be in the domain. Hence, the range of the function should not be calculated when $x = 2$ and $x = -5$.

Arithmetic Operations of Function

Examples

1. If $f(x) = 2x^2 + 1$, find:

a. $f(-2)$

b. $f(5)$

c. $f(0.3)$

d. $f(x - 1)$

e. $f(2x + 3)$

Solutions

a. $f(x) = 2x^2 + 1$

In order to obtain f(−2), we simply substitute −2 for x in the given function.

Hence, $f(x) = 2x^2 + 1$

$\quad f(-2) = 2(-2)^2 + 1$

$\qquad\quad = 2(4) + 1 = 8 + 1$

$\qquad\quad = 9$

Therefore, f(−2) = 9

b. $f(x) = 2x^2 + 1$

$\quad f(5) = 2(5)^2 + 1$

$\qquad\quad = 2(25) + 1 = 50 + 1$

$\qquad\quad = 51$

Therefore, f(5) = 51

c. $f(x) = 2x^2 + 1$

$\quad f(0.3) = 2(0.3)^2 + 1$

$\qquad\quad = 2(0.09) + 1$

$\qquad\quad = 0.18 + 1$

$\qquad\quad = 1.18$

Therefore, f(0.3) = 1.18

d. $f(x) = 2x^2 + 1$

$\quad f(x - 1) = 2(x - 1)^2 + 1$

$\qquad\quad = 2(x - 1)(x - 1) + 1$

$\qquad\quad = 2(x^2 - x - x + 1) + 1$

$\qquad\quad = 2(x^2 - 2x + 1) + 1$

$\qquad\quad = 2x^2 - 4x + 2 + 1$

$\qquad\quad = 2x^2 - 4x + 3$

Therefore, $f(x - 1) = 2x^2 - 4x + 3$

e. $f(x) = 2x^2 + 1$

    $f(2x + 3) = 2(2x + 3)^2 + 1$

              $= 2(2x + 3)(2x + 3) + 1$

              $= 2(4x^2 + 6x + 6x + 9) + 1$

              $= 2(4x^2 + 12x + 9) + 1$

              $= 8x^2 + 24x + 18 + 1$

              $= 8x^2 + 24x + 19$

Therefore, $f(2x + 3) = 8x^2 + 24x + 19$

2. A function f is defined by $f(x) = 5x - 8$

a. Find $f(\frac{1}{2})$

b. If $f(3m - 1) = 14$, find the value of m^2

c. Find $5f(-2)$

Solution

a. $f(x) = 5x - 8$

    $f(\frac{1}{2}) = 5(\frac{1}{2}) - 8$

        $= \frac{5}{2} - 8 = \frac{5 - 16}{2}$

          $= \frac{-11}{2}$

Therefore, $f(\frac{1}{2}) = -5\frac{1}{2}$

b. $f(x) = 5x - 8$

    $f(3m - 1) = 5(3m - 1) - 8$

           $14 = 15m - 5 - 8$ (Note that $f(3m - 1) = 14$)

           $14 = 15m - 13$

      $14 + 13 = 15m$

          $27 = 15m$

            $m = \frac{27}{15}$

            $m = \frac{9}{5}$

Therefore, $m^2 = (\frac{9}{5})^2$

          $m^2 = \frac{81}{25}$

c. $f(x) = 5x - 8$

    $f(-2) = 5(-2) - 8$

$$= -10 - 8$$
$$f(-2) = -18$$

Therefore, 5f(−2) is obtained by simply multiplying f(−2) by 5.

Therefore, 5f(−2) = 5(−18)
$$= -90$$

3. If $f(x) = 2x + 5$ and $g(x) = x - 2$, find:
a. $f(x) + g(x)$
b. $g(x) - 2f(x)$
c. $f(x) \times g(x)$
d. $f(2) \div g(5)$
e. $h(x) = f(x - 1) - g(-3)$
f. $h(-1)$

Solutions

a. $f(x) + g(x)$

This is the addition of the two functions. It is given as follows:
$$f(x) + g(x) = 2x + 5 + x - 2$$
$$= 3x + 3$$

b. $g(x) - 2f(x)$

In this case, 2f(x) has to be evaluated first. The overall solution is given by:
$$g(x) - 2f(x) = x - 2 - [2(2x + 5)]$$
$$= x - 2 - (4x + 10)$$
$$= x - 2 - 4x - 10$$
$$= -3x - 12$$

c. $f(x) \times g(x) = (2x + 5)(x - 2)$
$$= 2x^2 - 4x + 5x - 10$$
$$= 2x^2 + x - 10$$

d. $f(2) \div g(5)$

Let us determine f(2) and g(5) separately as follows:
$$f(x) = 2x + 5$$
$$f(2) = 2(2) + 5$$
$$= 4 + 5$$
$$= 9$$
$$g(x) = x - 2$$

$$g(5) = 5 - 2$$
$$= 3$$
$$\therefore \ f(2) \div g(5) = \frac{9}{3}$$
$$= 3$$

e. $h(x) = f(x - 1) - g(-3)$

Let us find $f(x - 1)$ as follows:

$\quad f(x) = 2x + 5$

Hence, $f(x - 1) = 2(x - 1) + 5$
$$= 2x - 2 + 5$$
$\quad f(x - 1) = 2x + 3$

Also, let us find $g(-3)$ as follows:

$\quad g(x) = x - 2$

$\quad g(-3) = -3 - 2$

$\quad g(-3) = -5$

$\therefore \ h(x) = f(x - 1) - g(-3)$
$$= 2x + 3 - (-5) \qquad \text{(Since } f(x - 1) = 2x + 3 \text{ and } g(-3) = -5)$$
$$= 2x + 3 + 5$$
$\quad h(x) = 2x + 8$

f. $h(x) = 2x + 8$

$\quad h(-1) = 2(-1) + 8$
$$= -2 + 8$$
$\quad h(-1) = 6$

Composing Functions

Chains of functions can be obtained when two or more functions combine together. For example if $f(x) = 2x$ and $g(x) = x - 1$, then a third function such as $h(x) = g[f(x)]$ can be obtained by the combination of $f(x)$ and $g(x)$. It is written as $h(x) = g \text{ o } f(x)$, and is read as h of x equals g of f of x. h is said to be a function of function. Note that any letter can be used to represent a function.

Examples

1. If $a(x) = 2x - 1$ and $b(x) = -3x$, find

a. $f(x) = a[b(x)]$

b. $g(x) = b[a(x)]$

Solutions

a. $f(x) = a[b(x)]$

This can also be written as: $f(x) = a \circ b(x)$

In order to find $f(x)$, we simply substitute $b(x)$ (i.e. $-3x$) for x in $a(x)$. This means to put $x = -3x$ in $a(x)$. This gives:

$\qquad f(x) = a[b(x)]$

$\qquad\qquad = a(-3x) \qquad$ (Since $b(x) = -3x$)

But, $a(x) = 2x - 1$

$\therefore\ a(-3x) = 2(-3x) - 1$

$\qquad\qquad = -6x - 1$

Hence, $f(x) = -6x - 1 \qquad$ [Since $f(x) = a(-3x)$]

b. $g(x) = b[a(x)]$

In order to find $g(x)$, we simply substitute $a(x)$ (i.e. $2x - 1$) for x in $b(x)$. This means we put $x = 2x - 1$ in $b(x)$. This gives:

$\qquad g(x) = b[a(x)]$

$\qquad\qquad = b(2x - 1) \qquad$ (Since $a(x) = 2x - 1$)

But, $b(x) = -3x$

$\therefore\ b(2x - 1) = -3(2x - 1)$

$\qquad\qquad = -6x + 3$

Hence, $g(x) = -6x + 3 \qquad$ [Since $g(x) = b(2x - 1)$]

2. If $a(x) = x^2$, $b(x) = 5x$ and $c(x) = x - 1$, find:

a. $f(x) = b[a[c(x)]]$

b. $g(x) = c[b[c(x)]]$

c. $h(x) = a[b[b(x)]]$

Solution

a. $f(x) = b[a[c(x)]]$

This can also be written as $b \circ a \circ c(x)$.

In order to find $f(x)$, we start from the innermost bracket. Hence, let us first find $a[c(x)]$. This simply means to substitute $c(x)$ in $a(x)$

$c(x) = x - 1$ and $a(x) = x^2$.

$\therefore\ a[c(x)] = a(x - 1) \qquad$ (Since $c(x) = x - 1$)

But, $\quad a(x) = x^2$

Hence, $a(x - 1) = (x - 1)^2$

$\qquad\qquad\qquad = (x - 1)(x - 1)$

$\qquad\qquad\qquad = x^2 - x - x + 1$

$\qquad a(x - 1) = x^2 - 2x + 1$

$\therefore\ a[c(x)] = x^2 - 2x + 1 \quad$ (Since $a[c(x)] = a(x - 1)$)

94

Let us now determine the final output i.e. $b[a[c(x)]]$. This means to substitute $a[c(x)]$ in $b(x)$

$\quad$ $a[c(x)] = x^2 - 2x + 1$ and $b(x) = 5x$

Hence, $b[a[c(x)]] = b(x^2 - 2x + 1)$ $\quad$ (Since $a[c(x)] = x^2 - 2x + 1$)

But, $b(x) = 5x$

Therefore, $b(x^2 - 2x + 1) = 5(x^2 - 2x + 1)$

$\quad\quad\quad\quad\quad\quad\quad = 5x^2 - 10x + 5$

Hence, $b[a[c(x)]] = 5x^2 - 10x + 5$ $\quad$ [Since $b[a[c(x)]] = b(x^2 - 2x + 1)$]

Therefore, $f(x) = 5x^2 - 10x + 5$ $\quad$ (Note that $f(x) = b[a[c(x)]]$)

b. $g(x) = c[b[c(x)]]$

Let us first find $b[c(x)]$. This simply means to substitute $c(x)$ in $b(x)$

$\quad$ $c(x) = x - 1$ and $b(x) = 5x$

∴ $\quad b[c(x)] = b(x - 1)$ $\quad$ (Since $c(x) = x - 1$)

But, $\quad b(x) = 5x$

Hence, $b(x - 1) = 5(x - 1)$

$\quad\quad$ $b(x - 1) = 5x - 5$

∴ $\quad\quad b[c(x)] = 5x - 5$ $\quad$ (Since $b[c(x)] = b(x - 1)$)

Let us now determine the final output i.e. $c[b[c(x)]]$. This means to substitute $b[c(x)]$ in $c(x)$

Hence, $\quad b[c(x)] = 5x - 5$ and $c(x) = x - 1$

Hence, $c[b[c(x)]] = c(5x - 5)$ $\quad$ [Since $b[c(x)] = 5x - 5$)]

But, $\quad c(x) = x - 1$

Therefore, $c(5x - 5) = (5x - 5) - 1$

$\quad\quad\quad\quad\quad c(5x - 5) = 5x - 6$

Hence, $c[b[c(x)]] = 5x - 6$ $\quad$ [Since $c[b[c(x)]] = c(5x - 5)$]

Therefore, $g(x) = 5x - 6$ $\quad$ (Note that $g(x) = c[b[c(x)]]$)

c. $h(x) = a[b[b(x)]]$

Note that $a[b[b(x)]]$ is interpreted as a of b of b of x or a o b o b(x).

Let us first find $b[b(x)]$. This simply means to substitute $b(x)$ in $b(x)$

$\quad$ $b(x) = 5x$

∴ $\quad b[b(x)] = b(5x)$

But, $b(x) = 5x$

Hence, $b(5x) = 5(5x)$

$\quad\quad$ $b(5x) = 25x$

∴ $\quad\quad b[b(x)] = 25x$

Let us now find $a[b[b(x)]]$. This means to substitute $b[b(x)]$ in $a(x)$

Recall that: $\quad b[b(x)] = 25x$ and $a(x) = x^2$

Hence, $a[b[b(x)]] = a(25x)$ $\quad$ [Since $b[b(x)] = 25x$]

But, $a(x) = x^2$

Therefore, $a(25x) = (25x)^2$

$\qquad a(25x) = 625x^2$

Hence, $\quad a[b[b(x)]] = 625x^2 \qquad$ [Since $a[b[b(x)]] = a(25x)$]

Therefore, $h(x) = 625x^2 \qquad$ (Since $h(x) = a[b[b(x)]]$)

Continuous and Discontinuous Functions

If a graph of a function is drawn without having to take ones hand off the graph paper, then the function is a continuous function. The graph of a continuous function has no sudden jump or break. Sine and cosine functions are continuous functions.

Some graphs of some functions make jumps at a point or some points in the interval. Such functions are called discontinuous functions. $y = \tan x$ is a discontinuous function.

Even Functions

A function $f(x)$ is said to be even if $f(x) = f(-x)$ for all values of x. The graphs of even functions always have their line of symmetry as the y–axis. This means that the vertical line $x = 0$ divide the graph into two equal parts.

$f(x) = x^2$ and $f(x) = \cos x$ are examples of even functions.

Odd Functions

A function $f(x)$ is said to be odd if $f(-x) = -f(x)$ for all values of x. Graphs of odd functions usually pass through the origin as a point of symmetry. This means that the origin, $(0, 0)$ divides the graph into two equal parts. $F(x) = x^3$ and $f(x) = \sin x$ are examples of odd functions.

Note that some functions are neither even nor odd.

Examples

1. Classify the following into even and odd functions:

a. $f(x) = \tan x$

b. $f(x) = x^4$

Solution

a. $f(x) = \tan x$

Let us take a value of x such as $60°$ and use it to test the function as follows:

$\qquad F(60°) = \tan 60$

$\qquad F(60°) = 1.732$

Now, find $f(-60°)$ to see if the value obtained will be the same as that of $f(60°)$.

$\qquad f(-60°) = \tan -60$

$\qquad f(-60°) = -1.732$

Comparing the two results [i.e. f(60°) and f(–60°)] shows that f(60°) = 1.732, while f(–60°) = – 1.732. Hence the function is an odd function since:

F(–x) = –f(x), i.e. f(–60°) = –f(60°)

Note that once the two values obtained are the same size, but different signs, then the function is an odd function.

b. $f(x) = x^4$

Let us take a value of x such as 2.

Hence, $f(2) = 2^4$

$f(2) = 16$

Now use the negative value of x i.e. –2. This gives:

$f(–2) = –2^4$

$F(–2) = 16$

Therefore, $f(2) = f(–2) = 16$

Since the two values obtained are equal and of the same sign, then the function is an even function.

2. Classify the following into even and odd functions:

a. $3x^5$

b. $\cos^3 x$

c. $4x – 1$

Solutions

a. $3x^5$

Let us use $x = 1$

Hence, $3x^5 = 3(1)^5$

$= 3$

When $x = –1$, we have:

$3x^5 = 3(–1)^5$

$= 3 \times (–1) = –3$ (Note that $(–1)^5 = –1$)

Since the values are the same but have opposite signs, then the function is an odd function.

b. $\cos^3 x$

Let $x = 60°$

$\cos^3 x = (\cos x)^3$

$= (\cos 60)^3$

$= (0.5)^3$

$= 0.125$

When $x = –60°$, we have:

$$\text{Cos}^3x = (\cos x)^3$$
$$= [\cos(-60)]^3$$
$$= (0.5)^3$$
$$= 0.125$$

The two values obtained are equal. Therefore, the function is an even function.

c. $4x - 1$

Let $x = 5$

$$4x - 1 = 4(5) - 1$$
$$= 20 - 1$$
$$= 19$$

When $x = -5$, we have

$$4x - 1 = 4(-5) - 1$$
$$= -20 - 1$$
$$= -21$$

The two values obtained are not equal. They are entirely different in values. Hence, the function is neither even nor odd.

Inverse of a Function

The inverse of a function f(x), is another function denoted by $f^{-1}(x)$ which reverses the function f(x). For an inverse $f^{-1}(x)$ that exists, the following is true:

$$f[\,f^{-1}(x)] = f^{-1}[f(x)] = x$$

The graph of $f^{-1}(x)$ will be a reflection of f(x) in the line y = x.

In order to find the inverse of a function, interchange the positions of x and y, and then make y the subject of the formula.

Examples

1. Find the inverse of the following functions:
a. f(x) = $5x - 1$
b. f(x) = $2x^2 + 3$
c. f(x) = $5x^3$

Solution

a. f(x) = $5x - 1$

Express the function as y = f(x). This gives:

$$y = 5x - 1$$

Interchange the positions of x and y. This gives:

$$x = 5y - 1$$

Now make y the subject of the formula.

$x = 5y - 1$

$x + 1 = 5y$

$y = \dfrac{x + 1}{5}$ (After dividing both sides by 5)

Therefore, the inverse of $f(x)$ is:

$f^{-1}(x) = \dfrac{x + 1}{5}$ [Simply by replacing y with $f^{-1}(x)$]

b. $f(x) = 2x^2 + 3$

As solved in (a) above, only three steps are involved. They are:

1. equate the function to y
2. interchange the positions of x and y
3. make y the subject of the formula to obtain the inverse of the function.

Note that step 1 is not needed if the function is already equated to y.

Let us now continue as follows:

$f(x) = 2x^2 + 3$

$y = 2x^2 + 3$

$x = 2y^2 + 3$

$x - 3 = 2y^2$

$y^2 = \dfrac{x - 3}{2}$

$y = \sqrt{\dfrac{x - 3}{2}}$ (This is the required inverse)

Therefore, the inverse of $f(x)$ is:

$f^{-1}(x) = \sqrt{\dfrac{x - 3}{2}}$

c. $f(x) = 5x^3$

$y = 5x^3$

$x = 5y^3$

$\dfrac{x}{5} = y^3$

$y = \sqrt[3]{\dfrac{x}{5}}$ (This is the required inverse)

Therefore, the inverse of $f(x)$ is:

$f^{-1}(x) = \sqrt[3]{\dfrac{x}{5}}$

2. Find the inverse of the following functions:

a. $f(x) = \left(\dfrac{2x - 1}{5}\right)^2$

b. $f(x) = (\dfrac{x+1}{3x-4})^{\frac{2}{3}}$

Solutions

a. $f(x) = (\dfrac{2x-1}{5})^2$

$y = (\dfrac{2x-1}{5})^2$

$x = (\dfrac{2y-1}{5})^2$

Taking the square root of both sides gives:

$\sqrt{x} = \sqrt{(\dfrac{2y-1}{5})^2}$

$\sqrt{x} = \dfrac{2y-1}{5}$

Cross multiply to obtain:

$2y - 1 = 5\sqrt{x}$

$2y = 5\sqrt{x} + 1$

$y = \dfrac{5\sqrt{x}+1}{2}$

Therefore, the inverse of $f(x)$ is:

$f^{-1}(x) = \dfrac{5\sqrt{x}+1}{2}$

b. $f(x) = (\dfrac{x+1}{3x-4})^{\frac{2}{3}}$

$y = (\dfrac{x+1}{3x-4})^{\frac{2}{3}}$

$x = (\dfrac{y+1}{3y-4})^{\frac{2}{3}}$

Raise both sides of the equation to a power of $\dfrac{3}{2}$ i.e. the inverse of $\dfrac{2}{3}$ in order to make the power of $(\dfrac{y+1}{3y-4})^{\frac{2}{3}}$ to become 1. This gives:

$x^{\frac{3}{2}} = ((\dfrac{y+1}{3y-4})^{\frac{2}{3}})^{\frac{3}{2}}$

$x^{\frac{3}{2}} = (\dfrac{y+1}{3y-4})^1$ (Note that the powers are multiplied to give 1, i.e. $\dfrac{3}{2} \times \dfrac{2}{3} = 1$)

$(\sqrt{x})^3 = \dfrac{y+1}{3y-4}$ (Note that from indices, $a^{\frac{3}{2}} = (\sqrt{a})^3 = \sqrt{a^3}$

$\sqrt{x^3} = \dfrac{y+1}{3y-4}$

Cross multiply to obtain:

$$\sqrt{x^3}(3y - 4) = y + 1$$

$$3y\sqrt{x^3} - 4\sqrt{x^3} = y + 1$$

Collect terms in y to obtain:

$$3y\sqrt{x^3} - y = 1 + 4\sqrt{x^3}$$

Factorizing the left hand side gives:

$$y(3\sqrt{x^3} - 1) = 1 + 4\sqrt{x^3}$$

Divide both sides by $3\sqrt{x^3} - 1$ to obtain y as follows:

$$y = \frac{1 + 4\sqrt{x^3}}{3\sqrt{x^3} - 1} \qquad \text{(This is the required inverse)}$$

Therefore, the inverse of f(x) is:

$$f^{-1}(x) = \frac{1 + 4\sqrt{x^3}}{3\sqrt{x^3} - 1}$$

3. If $f(x) = \dfrac{3x + 1}{x + 5}$, find:

a. $f^{-1}(x)$

b. $f^{-1}(-2)$

Solution

a. $f(x) = \dfrac{3x + 1}{x + 5}$

$$y = \frac{3x + 1}{x + 5}$$

$$x = \frac{3y + 1}{y + 5}$$

Cross multiply to obtain:

$$3y + 1 = x(y + 5)$$

$$3y + 1 = xy + 5x$$

Collect terms in y on the left hand side of the equation.

$$3y - xy = 5x - 1$$

Factorize the left hand side to obtain:

$$y(3 - x) = 5x - 1$$

Dividing both sides of the equation by $3 - x$ gives y as follows:

$$y = \frac{5x - 1}{3 - x}$$

Or, $y = \dfrac{-(1 - 5x)}{-(x - 3)}$ (After factorizing by taking −1 as a factor)

$$y = \frac{1 - 5x}{x - 3} \qquad \text{(After the negative signs cancel out)}$$

Hence, $y = \dfrac{5x - 1}{3 - x}$ or $y = \dfrac{1 - 5x}{x - 3}$

Therefore, the inverse of $f(x)$ is:

$$f^{-1}(x) = \frac{5x - 1}{3 - x} \quad \text{or} \quad f^{-1}(x) = \frac{1 - 5x}{x - 3} \qquad \text{(Note that any two of the inverses is correct)}$$

b. $f^{-1}(x) = \dfrac{5x - 1}{3 - x}$

$$f^{-1}(-2) = \frac{5(-2) - 1}{3 - (-2)}$$

$$= \frac{-10 - 1}{3 + 2}$$

$$= \frac{-11}{5}$$

$$\therefore \quad f^{-1}(-2) = -2\frac{1}{5}$$

Further Worked Examples on Functions

1. If $f(x - 1) = x^2 - 4x + 3$, find:

a. $f(x)$

b. $f(3)$

Solution

a. $f(x - 1) = x^2 - 4x + 3$

Let $y = x - 1$

Make x the subject of the formula. This gives:

$$x = y + 1$$

We now substitute $y + 1$ for x in the function above. This gives:

$$F(y) = f(x - 1) = x^2 - 4x + 3$$
$$= (y + 1)^2 - 4(y + 1) + 3$$
$$= (y + 1)(y + 1) - 4(y + 1) + 3$$
$$= y^2 + 2y + 1 - 4y - 4 + 3$$
$$F(y) = y^2 - 2y$$

Now, put $y = x$ into $f(y)$ in order to obtain $f(x)$ as follows:

$$F(x) = x^2 - 2x$$

b. $F(x) = x^2 - 2x$

$$f(3) = (3)^2 - 2(3)$$
$$= 9 - 6$$
$$= 3$$

2. If $f(2x + 1) = 5x - 3$, find $f(x + 3)$

Solution

$$f(2x + 1) = 5x - 3$$

Let y = 2x + 1

Make x the subject of the formula. This gives:

$$x = \frac{y - 1}{2}$$

We now substitute $\frac{y - 1}{2}$ for x in the function above. This gives:

$$F(y) = f(2x + 1) = 5x - 3$$

$$= 5\left(\frac{y - 1}{2}\right) - 3$$

$$= \frac{5y - 5}{2} - 3$$

$$= \frac{5y - 5 - 6}{2}$$

$$F(y) = \frac{5y - 11}{2}$$

This can now be written as a function of x by substituting x for y

$$F(x) = \frac{5x - 11}{2}$$

Substitute x + 3 for x in order to obtain f(x + 3) as follows:

$$F(x) = \frac{5x - 11}{2}$$

$$F(x + 3) = \frac{5(x + 3) - 11}{2}$$

$$= \frac{5x + 15 - 11}{2}$$

$$F(x + 3) = \frac{5x + 4}{2}$$

3. Given that f(x + 3) = $x^2 - 7$, find f(−1)

Solution

$$f(x + 3) = x^2 - 7$$

We are going to use a more direct method which is different from that used in examples 1 and 2 above.

$$f(x + 3) = x^2 - 7$$

In order to find f(−1), we simply equate the two terms in bracket and make x the subject of the formula. Hence, we have:

$$f(x + 3) = f(-1)$$

$$(x + 3) = -1$$

$$x = -1 - 3$$

$$x = -4$$

We now substitute −4 for x in the original function in order to obtain f(−1). This gives:

F(−1) = f(−4) in the function of f(x + 3)

Hence, $f(x + 3) = x^2 - 7$

$\quad$ F(−4) = (−4)^2 − 7

$\quad\quad$ = 16 − 7 = 9

Therefore, f(−1) = 9 $\quad\quad$ [Since f(−1) = f(−4) in the function of f(x + 3)]

4. If f(x + 2) = 2x − 5, find:

a. f(x)

b. f^{-1}(x)

c. f(−3)

d. f^{-1}(−3)

e. f(x − 3)

Solution

a. $\quad$ f(x + 2) = 2x − 5

In order to find f(x), equate the two terms in each of the brackets

$\quad\quad$ f(x + 2) = f(x)

Hence, x + 2 = x

$\quad\quad\quad$ x = x − 2

Be sure to make the x in the first bracket i.e. from the original function, to be the subject of the formula.

$\quad\quad$ Hence f(x) = f(x − 2) in the function of f(x + 2)

Now substitute x − 2 for x in the given function.

$\quad\quad$ F(x + 2) = 2x − 5

$\quad\quad$ F(x) = f(x − 2) = 2(x − 2) − 5 $\quad$ (x − 2 has been substituted for x in the function)

$\quad\quad\quad$ = 2x − 4 − 5

$\quad\quad$ F(x) = 2x − 9

b. $\quad$ F(x) = 2x − 9

Let us find f^{-1}(x) i.e. the inverse of f(x). This is done as follows:

$\quad\quad$ y = 2x − 9

$\quad\quad$ x = 2y − 9

$\quad\quad$ x + 9 = 2y

$\quad\quad$ $y = \dfrac{x + 9}{2}$

Therefore, $f^{-1}(x) = \dfrac{x + 9}{2}$

c. $\quad$ F(x) = 2x − 9

$$F(-3) = 2(-3) - 9$$
$$= -6 - 9$$
$$F(-3) = -15$$

d. $f^{-1}(x) = \dfrac{x + 9}{2}$

$f^{-1}(-3) = \dfrac{-3 + 9}{2}$

$= \dfrac{6}{2}$

$f^{-1}(-3) = 3$

e. $F(x) = 2x - 9$

$F(x - 3) = 2(x - 3) - 9$

$= 2x - 6 - 9$

$F(x - 3) = 2x - 15$

5. If $f(3x - 1) = 6x + 5$, find:

a. $f(2x + 5)$

b. $f(x^2)$

<u>Solution</u>

a. $f(3x - 1) = 6x + 5$

$f(2x + 5)$ is obtained as follows:

$3x - 1 = 2x + 5$ (By equating only terms in the brackets)

$3x = 2x + 5 + 1$

$3x = 2x + 6$

$x = \dfrac{2x + 6}{3}$

Hence, $f(2x + 5) = f(\dfrac{2x + 6}{3})$ in the function of $f(3x - 1)$

Therefore substitute $\dfrac{2x + 6}{3}$ for x in the original function above. This gives:

$f(3x - 1) = 6x + 5$

$f(2x + 5) = f(\dfrac{2x + 6}{3}) = 6(\dfrac{2x + 6}{3}) + 5$

$= 2(2x + 6) + 5$ (Note that $6 \div 3 = 2$)

$= 4x + 12 + 5$

$F(2x + 5) = 4x + 17$

b. $f(3x - 1) = 6x + 5$

105

$F(x^2)$ will be obtained from $f(3x-1)$ as follows

$$3x - 1 = x^2 \quad \text{(By equating only terms in the brackets)}$$
$$3x = x^2 + 1$$
$$x = \frac{x^2 + 1}{3}$$

Hence, $f(x^2) = f(\frac{x^2 + 1}{3})$ in the function of $f(3x-1)$

Therefore substitute $\frac{x^2 + 1}{3}$ for x in the original function above. This gives:

$$f(3x - 1) = 6x + 5$$
$$f(x^2) = f(\frac{x^2 + 1}{3}) = 6(\frac{x^2 + 1}{3}) + 5$$
$$= 2(x^2 + 1) + 5$$
$$= 2x^2 + 2 + 5$$
$$F(x^2) = 2x^2 + 7$$

Exercise 8

1. Determine if each of the following is a function or not.

a. $y = x^{\frac{1}{2}} + 5$

b. $y = 2x^3 - \sqrt[4]{x}$

c. $y = \sqrt{5 - x^5}$

d. $y = \dfrac{3}{5x - 2}$

2. If $f(x) = 3x^3 - 2$, find:

a. $f(-1)$

b. $f(2)$

c. $f(0.5)$

d. $f(3x - 2)$

e. $f(x + 3)$

3. A function f is defined by $f(x) = 2x - 3$

a. Find $f(\frac{1}{5})$

b. If $f(2p + 3) = 10$, find the value of p

c. Find $9f(-\frac{1}{5})$

4. If $f(x) = x - 3$ and $g(x) = 2x + 1$, find:

a. $f(x) + g(x)$

b. $g(x) - 5f(x)$

c. $f(x) \times g(x)$

d. $f(-1) \div g(2)$

e. $h(x) = f(x - 2) - g(5)$

f. $h(2)$

5. If $a(x) = x - 2$ and $b(x) = 2x$, find

a. $f(x) = b[a(x)]$

b. $g(x) = a[b(x)]$

6. If $a(x) = 2x^2$, $b(x) = 3x$ and $c(x) = 2x + 1$, find:

a. $f(x) = a[b[c(x)]]$

b. $g(x) = b[b[a(x)]]$

c. $h(x) = c[a[b(x)]]$

7. Classify the following into even and odd functions:

a. $f(x) = \cos x$

b. $f(x) = 2x^3$

8. Classify the following into even and odd functions:

a. $4x^2$

b. $\cos^2 3x$

c. $9x - 4$

d. $\sin^2 x$

9. Find the inverse of the following functions:

a. $f(x) = 5x - 1$

b. $f(x) = 9x^3 + 2$

c. $f(x) = 2x^5$

10. Find the inverse of the following functions:

a. $f(x) = (\dfrac{x - 1}{2})^3$

b. $f(x) = (\dfrac{2x + 1}{5x - 2})^{\frac{1}{2}}$

11. If $f(x) = \dfrac{5x - 1}{2x + 1}$, find:

a. $f^{-1}(x)$

b. $f^{-1}(-1)$

12. If $f(x - 2) = 2x^2 - 5x - 7$, find:

a. $f(x)$

b. $f(-2)$

13. If $f(5x - 1) = 2x + 7$, find $f(x - 2)$

14. Given that $f(x + 4) = x^2 - 2$, find $f(-3)$

15. If $f(2x + 1) = 4x - 3$, find:

a. $f(x)$

b. $f^{-1}(x)$

c. $f(-2)$

d. $f^{-1}(-1)$

e. $f(2x - 1)$

16. If $f(x - 3) = 7x + 2$, find:

a. $f(x - 5)$

b. $f(2x^2)$

CHAPTER 9
POLYNOMIALS

A polynomial is an expression which is a sum of terms containing a variable or variables whose power starts from one and above.

The highest power of a variable in a polynomial is called the degree of the polynomial.

Addition and Subtraction of Polynomials

When adding or subtracting polynomials, the like terms are added or subtracted as the case may be. Note that like terms are terms whose variables have the same power.

Examples

1. If $A = 2x^3 + 7x^2 - 5$, $B = 5x^3 - 11$, and $C = x^3 + 2x^2 + 5x + 3$, find:

a. $A + B$

b. $C - B$

c. $2B + A$

d. $A - 2C + 3B$

e. $A - B - C$

<u>Solutions</u>

a. $A + B = (2x^3 + 7x^2 - 5) + (5x^3 - 11)$

$\qquad = 2x^3 + 5x^3 + 7x^2 - 5 - 11$ (Take note of how like terms are brought together)

$\qquad = 7x^3 + 7x^2 - 16$

Take note of the arrangement of terms in ascending order of powers of the variables.

b. $C - B = (x^3 + 2x^2 + 5x + 3) - (5x^3 - 11)$

$\qquad = x^3 + 2x^2 + 5x + 3 - 5x^3 + 11$

$\qquad = x^3 - 5x^3 + 2x^2 + 5x + 3 + 11$

$\qquad = -4x^3 + 2x^2 + 5x + 14$ (Note that x^3 also means $1x^3$)

c. $2B + A = 2(5x^3 - 11) + (2x^3 + 7x^2 - 5)$

$\qquad = 10x^3 - 22 + 2x^3 + 7x^2 - 5$

$\qquad = 10x^3 + 2x^3 + 7x^2 - 22 - 5$

$\qquad = 12x^3 + 7x^2 - 27$

d. $A - 2C + 3B = 2x^3 + 7x^2 - 5 - 2(x^3 + 2x^2 + 5x + 3) + 3(5x^3 - 11)$

$\qquad = 2x^3 + 7x^2 - 5 - 2x^3 - 4x^2 - 10x - 6 + 15x^3 - 33$

Take note of how a negative sign outside a bracket changes all the signs in the bracket.

$\qquad = 2x^3 - 2x^3 + 15x^3 + 7x^2 - 4x^2 - 10x - 5 - 6 - 33$

$\qquad = 15x^3 + 3x^2 - 10x - 44$

e. $A - B - C = (2x^3 + 7x^2 - 5) - (5x^3 - 11) - (x^3 + 2x^2 + 5x + 3)$
$= 2x^3 + 7x^2 - 5 - 5x^3 + 11 - x^3 - 2x^2 - 5x - 3$
$= 2x^3 - 5x^3 - x^3 + 7x^2 - 2x^2 - 5x - 5 + 11 - 3$
$= -4x^3 + 5x^2 - 5x + 3$

2. If $f(x) = 2x^3 - x^2 + 6x - 5$, find:
a. $f(-2)$
b. $f(0)$

<u>Solutions</u>

a. $f(x) = 2x^3 - x^2 + 6x - 5$

In order to find $f(-2)$, simply substitute -2 for x in the given function. This gives:
$F(-2) = 2(-2)^3 - (-2)^2 + 6(-2) - 5$
$F(-2) = 2(-8) - (4) - 12 - 5$
$= -16 - 4 - 12 - 5$
$= -37$

b. $f(x) = 2x^3 - x^2 + 6x - 5$
$f(0) = 2(0)^3 - (0)^2 + 6(0) - 5$
$= 0 - 0 + 0 - 5$
$= -5$

Multiplication of Polynomials

When multiplying polynomials, remember to add the powers of the variable according to the multiplication law of indices.

Examples

1. If $A = 2x^3 - 5x^2 + 3x$ and $B = 3x^3 + x^2 - 7x - 5$, find AB.

<u>Solution</u>

METHOD 1

In order to carry out this multiplication, use each term in A to multiply all the terms in B. The use of bracket in doing this is necessary. Also, remember to carry the sign of each term in A.

$\therefore$ AB $= (2x^3 - 5x^2 + 3x)(3x^3 + x^2 - 7x - 5)$
$= 2x^3(3x^3 + x^2 - 7x - 5) - 5x^2(3x^3 + x^2 - 7x - 5) + 3x(3x^3 + x^2 - 7x - 5)$
$= 6x^6 + 2x^5 - 14x^4 - 10x^3 - 15x^5 - 5x^4 + 35x^3 + 25x^2 + 9x^4 + 3x^3 - 21x^2 - 15x$

Take note of the addition of the powers of x. We now bring like terms together, i.e. terms having the same powers of x. After that we add/subtract the like terms. This gives:
AB $= 6x^6 + 2x^5 - 15x^5 - 14x^4 - 5x^4 + 9x^4 - 10x^3 + 35x^3 + 3x^3 + 25x^2 - 21x^2 - 15x$

$$AB = 6x^6 - 13x^5 - 10x^4 + 28x^3 + 4x^2 - 15x$$

METHOD 2

In order to carry out this method, we arrange the terms in columns with the polynomial having the higher number of terms above the one having lower number of terms. B has higher number of terms, so we place B above A. Note that AB = BA. Also, arrange like terms above each other when carrying out the initial arrangement and during the multiplication of terms. Note that we use each term in the lower column to multiply all the terms in the higher column by starting from the right hand side of the arrangement. This means that we work from the right hand side to the left hand side. It is similar to the way we carry out multiplication of numbers.

Let us now multiply A and B as follows:

$$
\begin{array}{r}
3x^3 + x^2 - 7x - 5 \\
2x^3 - 5x^2 + 3x \\
\hline
+ 9x^4 + 3x^3 - 21x^2 - 15x \\
- 15x^5 - 5x^4 + 35x^3 + 25x^2 \\
+ 6x^6 + 2x^5 - 14x^4 - 10x^3 \\
\hline
6x^6 - 13x^5 - 10x^4 + 28x^3 + 4x^2 - 15x
\end{array}
$$

2. If $M = x^3 - 8x^2 + 2x + 1$ and $N = 5x^3 + 2x^2 - 4x - 7$, find MN.

Solution

METHOD 1

$MN = (x^3 - 8x^2 + 2x + 1)(5x^3 + 2x^2 - 4x - 7)$

$= x^3(5x^3 + 2x^2 - 4x - 7) - 8x^2(5x^3 + 2x^2 - 4x - 7) + 2x(5x^3 + 2x^2 - 4x - 7) + 1(5x^3 + 2x^2 - 4x - 7)$

$= 5x^6 + 2x^5 - 4x^4 - 7x^3 - 40x^5 - 16x^4 + 32x^3 + 56x^2 + 10x^4 + 4x^3 - 8x^2 - 14x + 5x^3 + 2x^2 - 4x - 7$

$= 5x^6 + 2x^5 - 40x^5 - 4x^4 - 16x^4 + 10x^4 - 7x^3 + 32x^3 + 4x^3 + 5x^3 + 56x^2 - 8x^2 + 2x^2 - 14x - 4x - 7$

$MN = 5x^6 - 38x^5 - 10x^4 + 34x^3 + 50x^2 - 18x - 7$

METHOD 2

Ensure you multiply the signs of any two terms multiplied together. The working is as shown below.

$$
\begin{array}{r}
x^3 - 8x^2 + 2x + 1 \\
5x^3 + 2x^2 - 4x - 7 \\
\hline
- 7x^3 + 56x^2 - 14x - 7 \\
- 4x^4 + 32x^3 - 8x^2 - 4x \\
2x^5 - 16x^4 + 4x^3 + 2x^2 \\
+ 5x^6 - 40x^5 + 10x^4 + 5x^3 \\
\hline
5x^6 - 38x^5 - 10x^4 + 34x^3 + 50x^2 - 18x - 7
\end{array}
$$

3. Given that $f(x) = 2x^3 - 3$ and $g(x) = 3x^3 - 2x^2 + x - 5$, find $f(x) \cdot g(x)$

<u>Solution</u>

METHOD 1

$$\begin{aligned}
f(x) \times g(x) &= (2x^3 - 3)(3x^3 - 2x^2 + x - 5) \\
&= 2x^3(3x^3 - 2x^2 + x - 5) - 3(3x^3 - 2x^2 + x - 5) \\
&= 6x^6 - 4x^5 + 2x^4 - 10x^3 - 9x^3 + 6x^2 - 3x + 15 \\
&= 6x^6 - 4x^5 + 2x^4 - 19x^3 + 6x^2 - 3x + 15
\end{aligned}$$

METHOD 2

Since $g(x)$ has more terms, we will place $g(x)$ above $f(x)$. Also arrange like terms above each other as shown below

$$\begin{array}{r}
3x^3 - 2x^2 + x - 5 \\
\underline{2x^3 \qquad\qquad\quad - 3} \\
-9x^3 + 6x^2 - 3x + 15 \\
\underline{+\; 6x^6 - 4x^5 + 2x^4 - 10x^3 \qquad\qquad\qquad} \\
6x^6 - 4x^5 + 2x^4 - 19x^3 + 6x^2 - 3x + 15
\end{array}$$

Division of Polynomials

If we divide $2x^3 - 2x^2 + x - 5$ by $x - 5$, then $2x^3 - 2x^2 + x - 5$ is called dividend, while $x - 5$ is called the divisor. The result obtained after the division is called the quotient, while what is left at the end of the division is called the remainder.

Examples

1. Divide $2x^2 + 3x - 5$ by $x - 1$

<u>Solution</u>

The layout is as shown below. Note that 'like terms' are arranged in columns above each other when carrying out the division.

STEP 1: Divide the first term of the dividend by the first term of the divisor. This gives: $\dfrac{2x^2}{x} = 2x$.

Then write the $2x$ at the top of the division sign as shown below.

$$\begin{array}{r}
2x \qquad\quad \\
x - 1 \overline{)2x^2 + 3x - 5}
\end{array}$$

STEP 2: Use the $2x$ obtained above to multiply $x - 1$ and write your answer under the dividend. Note that $2x(x - 1)$ will give $2x^2 - 2x$. This is now written below the corresponding like terms of the dividend as shown below.

$$\begin{array}{r}
2x \qquad\quad \\
x - 1 \overline{)2x^2 + 3x - 5} \\
2x^2 - 2x \qquad\;
\end{array}$$

STEP 3: Subtract the like terms as arranged above. This means: $2x^2 - 2x^2 = 0x^2$, while $3x - (-2x)$ $= 3x + 2x = 5x$. We now write $5x$ under its corresponding column and ignore the zero under $2x^2$ since there is no need of writing $0x^2$. This is as shown below.

$$\begin{array}{r} 2x \phantom{{}+3x-5} \\ x-1\overline{)2x^2 + 3x - 5} \\ \underline{2x^2 - 2x} \\ 5x \end{array}$$

STEP 4: Bring down the next term of the dividend which is –5. This is as shown below.

$$\begin{array}{r} 2x \phantom{{}+3x-5} \\ x-1\overline{)2x^2 + 3x - 5} \\ \underline{2x^2 - 2x} \\ 5x - 5 \end{array}$$

STEP 5: After bringing down –5, we now have a new dividend of $5x - 5$. Use this new dividend to repeat step 1 to step 4 above. This means:

Step 1: $\dfrac{5x}{x}$ = 5. Write this as +5 at the top of the division sign.

Step 2: Use the 5 above to multiply $x - 1$. This gives $5x - 5$. Write this below the corresponding like terms of our new dividend which is also $5x - 5$.

Step 3: Subtract the like terms arranged above. This will give zero since $5x - 5$ subtracted from $5x - 5$ gives zero.

Step 4: When your subtraction in step 3 gives zero, and there is no more term to bring down from the original dividend, then you have arrived at your answer. These steps give the final solution as shown below:

$$\begin{array}{r} 2x + 5 \\ x-1\overline{)2x^2 + 3x - 5} \\ \underline{2x^2 - 2x} \\ 5x - 5 \\ \underline{5x - 5} \\ -\ -\ - \end{array}$$

$\therefore$ $(2x^2 + 3x - 5) \div (x - 1) = 2x + 5$

Check your answer by multiplying $(x - 1)$ by $(2x + 5)$. It will give $2x^2 + 3x - 5$, which is the dividend.

Notice that the steps from 1 to 4 above can be formulated into an acronym written as: DMSBd. D means divide, M means multiply, S means subtract, while Bd means bring down. This acronym can always remind you of the next step to take.

2. Divide $5x^3 + x^2 - 8x - 4$ by $x + 1$

<u>Solution</u>

The question can also be written as:

$$\frac{5x^3 + x^2 - 8x - 4}{x + 1}$$

The workings are explained below.

STEP 1: Divide the first term of the dividend by the first term of the divisor. This gives: $\frac{5x^3}{x} = 5x^2$. Then write the $5x^2$ at the top of the division sign as shown below.

$$\begin{array}{r} 5x^2 \\ x+1 \overline{)5x^3 + x^2 - 8x - 4} \end{array}$$

STEP 2: Use the $5x^2$ obtained above to multiply $x + 1$ and write your answer under the dividend. Note that $5x^2(x + 1)$ will give $5x^3 + 5x^2$. This is now written below the corresponding like terms of the dividend as shown below.

$$\begin{array}{r} 5x^2 \\ x+1 \overline{)5x^3 + x^2 - 8x - 4} \\ 5x^3 + 5x^2 \end{array}$$

STEP 3: Subtract the like terms as arranged above. This means: $5x^3 - 5x^3 = 0x^3$, while $x^2 - (+5x^2) = x^2 - 5x^2 = -4x^2$. We now write $-4x^2$ under its corresponding column and ignore the zero under $5x^3$ since there is no need of writing $0x^3$. This is as shown below.

$$\begin{array}{r} 5x^2 \\ x+1 \overline{)5x^3 + x^2 - 8x - 4} \\ \underline{5x^3 + 5x^2} \\ -4x^2 \end{array}$$

STEP 4: Bring down the next term of the dividend which is $-8x$. This is as shown below.

$$\begin{array}{r} 5x^2 \\ x+1 \overline{)5x^3 + x^2 - 8x - 4} \\ \underline{5x^3 + 5x^2} \\ -4x^2 - 8x \end{array}$$

STEP 5: After bringing down $-8x$, we now have a new dividend of $-4x^2 - 8x$ as shown above. Use this new dividend to repeat step 1 to step 4 above. This means:

Step 1: $\frac{-4x^2}{x} = -4x$. Write this at the top of the division sign as shown below.

Step 2: Use the $-4x$ above to multiply $x + 1$. This gives $-4x^2 - 4x$. Write this below the corresponding like terms of our new dividend.

Step 3: Subtract the like terms arranged from step 2 above. This will give $-4x$.

Step 4: Bring down the next term of the original dividend which is -4.

Step 5: After bringing down -4, we now have a new dividend of $-4x - 4$. Use this new dividend to repeat step 1 to step 4 above.

Now let us use the acronym DMSBd to complete the remaining part of the division as follows.

D: $-4x \div x = -4$ (This is written on the division sign)

M: $-4(x + 1) = -4x - 4$ (Write this under $-4x - 4$ which is our present dividend)

S: $-4x - 4 - (-4x - 4)$ will give zero.

Bd: There is nothing more to bring down. Hence our division is complete and we now have our final answer.

All the division processes explained above are as shown below.

$$
\begin{array}{r}
5x^2 - 4x - 4 \\
x + 1\overline{)5x^3 + x^2 - 8x - 4} \\
\underline{5x^3 + 5x^2} \\
-4x^2 - 8x \\
\underline{-4x^2 - 4x} \\
-4x - 4 \\
\underline{-4x - 4} \\
-\quad -\quad -
\end{array}
$$

$\therefore$ $(5x^3 + x^2 - 8x - 4) \div (x + 1) = 5x^2 - 4x - 4$

Check your answer by multiplying $(x + 1)$ by $(5x^2 - 4x - 4)$. It will give $5x^3 + x^2 - 8x - 4$, which is the dividend.

Points to note as a reminder:

D: During division, only the first term of the dividend is used to divide the first term of the divisor.

M: During multiplication, the answer obtained during division is used to multiply all the terms in the divisor.

S: During subtraction, only the like terms are subtracted. The sign of each term must be carried along with it. Be careful of the negative sign of terms, and the subtraction sign used to carry out the operation.

Bd: Bring down the next term along with its sign. When there is nothing else to bring down, then the division process has ended and the terms on the division sign become the answer.

3. Evaluate $\dfrac{x^3 - 7x - 6}{x - 3}$

Solution

A close look at the dividend shows that there is no term in x^2. Therefore, in order to avoid mistake in our division, it is advisable to include $0x^2$ at the right position in the dividend. Hence the question can be re–written as:

$$\frac{x^3 - 0x^2 - 7x - 6}{x - 3}$$

We now set out our division as shown below.

$$\begin{array}{r} x^2 + 3x + 2 \\ x - 3\overline{)x^3 + 0x^2 - 7x - 6} \\ \underline{x^3 - 3x^2} \\ 3x^2 - 7x \\ \underline{3x^2 - 9x} \\ 2x - 6 \\ \underline{2x - 6} \\ -\ -\ - \end{array}$$

WORKING

D: $\dfrac{x^3}{x} = x^2$ (This is written on the division sign)

M: $x^2(x - 3) = x^3 - 3x^2$ (Write this under $x^3 + 0x^2$)

S: $0x^2 - (-3x^2) = 0x^2 + 3x^2 = 3x^2$.

Bd: Bring down $-7x$ to obtain $3x^2 - 7x$ as the new dividend.

We now repeat the process using $3x^2 - 7x$

D: $\dfrac{3x^2}{x} = 3x$ (This is written as $+3x$ on the division sign)

M: $3x(x - 3) = 3x^2 - 9x$ (Write this under $3x^2 - 7x$)

S: $3x^2 - 7x - (3x^2 - 9x) = 2x$.

Bd: Bring down -6 to meet $2x$. This gives $2x - 6$ as the new dividend.

Finally, we repeat the process by using $2x - 6$ as dividend.

D: $\dfrac{2x}{x} = 2$ (This is written as $+2$ on the division sign)

M: $2(x - 3) = 2x - 6$ (Write this under $2x - 6$)

S: Their subtraction gives zero

Bd: Nothing more to bring down. Hence we have our answer as $x^2 + 3x + 2$.

Therefore, $\dfrac{x^3 - 7x - 6}{x - 3} = x^2 + 3x + 2$

4. Divide $2x^3 - 11x^2y + 3xy^2 + y^3$ by $2x - y$

<u>Solution</u>

$$\begin{array}{r} x^2 - 5xy - y^2 \\ 2x - y\overline{)2x^3 - 11x^2y + 3xy^2 + y^3} \\ \underline{2x^3 - x^2y} \\ -10x^2y + 3xy^2 \\ \underline{-10x^2y + 5xy^2} \\ -2xy^2 + y^3 \\ \underline{-2xy^2 + y^3} \\ -\ -\ - \end{array}$$

5. Simplify: $\dfrac{x^3 - y^3}{x - y}$

Solution

A careful look at the dividend shows that some terms are not present. That is, they are zero. The missing terms are x^2y and xy^2. One method of knowing the missing term is to raise the divisor to the highest power (degree) of the dividend. This means that if $(x - y)^3$ is evaluated, you will see the missing terms.

Hence we represent the missing terms by $0x^2y$ and $0xy^2$. We now carry out the division as shown below.

$$
\begin{array}{r}
x^2 + xy + y^2 \\
x - y \overline{)x^3 + 0x^2y + 0xy^2 - y^3} \\
\underline{x^3 - x^2y} \\
x^2y + 0xy^2 \\
\underline{x^2y - xy^2} \\
xy^2 - y^3 \\
\underline{xy^2 - y^3} \\
- \quad - \quad -
\end{array}
$$

Hence, $\dfrac{x^3 - y^3}{x - y} = x^2 + xy + y^2$

6. Find the quotient and the remainder when $4x^3 - 6x^2 + 8x - 5$ is divided by $2x + 1$

Solution

This is a case where we will have a remainder. In the course of our division, whenever we carry out a subtraction, and there is nothing else to bring down, whatever is left becomes the remainder. Let us now carry out the division as follows.

$$
\begin{array}{r}
2x^2 - 4x + 6 \\
2x + 1 \overline{)4x^3 - 6x^2 + 8x - 5} \\
\underline{4x^3 + 2x^2} \\
-8x^2 + 8x \\
\underline{-8x^2 - 4x} \\
12x - 5 \\
\underline{12x + 6} \\
-11
\end{array}
$$

There is nothing else left to bring down. This ends the division.

$\therefore \quad (4x^3 - 6x^2 + 8x - 5) \div (2x + 1) = 2x^2 - 4x + 6$, remainder -11.

Hence, the quotient is $2x^2 - 4x + 6$, while the remainder is -11.

This division can be written as: $\dfrac{4x^3 - 6x^2 + 8x - 5}{2x + 1} = 2x^2 - 4x + 6 - \dfrac{11}{2x + 1}$

7. Find the quotient and the remainder when $3x^3 - 7x^2 - x + 9$ is divided by $x - 5$

Solution

Let us carry out the division as follows.

$$
\begin{array}{r}
3x^2 + 8x + 39 \\
x - 5 \overline{)\,3x^3 - 7x^2 - x + 9} \\
\underline{3x^3 - 15x^2} \\
8x^2 - x \\
\underline{8x^2 - 40x} \\
39x + 9 \\
\underline{39x - 195} \\
204
\end{array}
$$

There is nothing else left to bring down. This ends the division.

$\therefore$ The quotient is $3x^2 + 8x + 39$, while the remainder is 204.

This division can be written as: $\dfrac{3x^3 - 7x^2 - x + 9}{x - 5} = 3x^2 + 8x + 39 - \dfrac{204}{x - 5}$

Zeros of Polynomials

A zero of a function f(x) is the root of the equation f(x) = 0.

Examples

Find the zeros of the following polynomial functions:

1. $f(x) = x^2 - 5x + 6$
2. $f(x) = 2x^2 + 5x - 3$

Solution

1. In order to find the zeros of the function, we equate the function to zero and solve for x as follows:

$$x^2 - 5x + 6 = 0$$
$$(x - 2)(x - 3) = 0$$
$\therefore \quad x = 2 \text{ or } x = 3$

Hence, the zeros of the function are 2 and 3.

2. $2x^2 + 5x - 3 = 0$

Solving this equation by factorization gives:

$$2x^2 - x + 6x - 3 = 0$$
$$x(2x - 1) + 3(2x - 1) = 0$$
$$(2x - 1)(x + 3)$$
$$\therefore \quad x = \frac{1}{2} \text{ or } x = -3$$

Hence, the zeros of the function are $\frac{1}{2}$ and -3.

The Factor Theorem

Let us solve the quadratic equation below by factorization.

$$x^2 - 8x - 20 = 0$$
$$(x - 10)(x + 2) = 0$$
$$\therefore \quad x = 10 \text{ or } x = -2$$

This shows that the factor $(x - 10)$ gives a root of 10, while the factor $(x + 2)$ gives a root of -2. This is what the factor theorem means.

Therefore the factor theorem state that:

If $x = a$ is a root of the equation f(x) = 0, then $(x - a)$ is a factor of f(x). This also means that f(a) = 0 (i.e. substituting 'a' for x in the function gives zero).

The Remainder Theorem

The remainder theorem states that if f(x) is divided by $(x - a)$, the remainder is equal to f(a).

Generally, we can say that, if f(x) is divided by $ax - b$, the remainder is equal to $f(\frac{b}{a})$.

Example

1. Factorize $x^3 - 2x^2 - x + 2$ and use it to solve the cubic equation $x^3 - 2x^2 - x + 2 = 0$

Solution

Let us represent the expression as a function of x.

$$F(x) = x^3 - 2x^2 - x + 2$$

We have to employ the method of trial and error to obtain one factor of f(x). Try numbers such as 1, -1, 2, -2, 3 etc.

$$F(x) = x^3 - 2x^2 - x + 2$$

If $x = 1$, then f(1) = $(1)^3 - 2(1)^2 - (1) + 2$

$$= 1 - 2 - 1 + 2$$
$$= 0$$

Since f(1) = 0, then $(x - 1)$ is a factor of f(x) (factor theorem). Note that f(1) = 0 means that $x = 1$, which can be rearranged to give $x - 1 = 0$ (by taking 1 to the left hand side of the equation). Hence we can see that $x - 1$ is a factor of f(x).

Let us now use the polynomial to divide $(x - 1)$ in order to obtain the other factors. This is shown below.

$$
\begin{array}{r}
x^2 - x - 2 \\
x - 1)\overline{x^3 - 2x^2 - x + 2} \\
\underline{x^3 - x^2} \\
-x^2 - x \\
\underline{-x^2 + x} \\
-2x + 2 \\
\underline{-2x + 2} \\
- \quad - \quad -
\end{array}
$$

Hence, $\dfrac{x^3 - 2x^2 - x + 2}{x - 1} = x^2 - x - 2$

Or, $x^3 - 2x^2 - x + 2 = (x - 1)(x^2 - x - 2)$

The quadratic part $x^2 - x - 2$ can be factorized to give:

$x^2 - x - 2 = (x - 2)(x + 1)$

$\therefore \quad x^3 - 2x^2 - x + 2 = (x - 1)(x - 2)(x + 1)$

Let us now use the factorized expression above to solve the given equation as follows:

$x^3 - 2x^2 - x + 2 = 0$

$(x - 1)(x - 2)(x + 1) = 0$

Hence, $x = 1$ or 2 or -1　　(When each bracket is equated to zero and solved)

2. Solve the equation $2x^3 + 13x^2 + 13x - 10 = 0$

<u>Solution</u>

Let $f(x) = 2x^3 + 13x^2 + 13x - 10$

Let us put numbers such as $-1, 1, -2, 2, -3$ etc, into the function in order to obtain one of the factors.

Hence, if $x = 1$, $f(1) = 2(1)^3 + 13(1)^2 + 13(1) - 10$

$= 2 + 13 + 13 - 10$

$= 18$

Therefore, $(x - 1)$ is not a factor

If $x = -1$, $f(-1) = 2(-1)^3 + 13(-1)^2 + 13(-1) - 10$

$= -2 + 13 - 13 - 10$

$= -12$

Therefore, $(x + 1)$ is not a factor. Note that from $x = -1$, we have $x + 1$ by taking -1 to the left hand side.

If $x = 2$, $f(2) = 2(2)^3 + 13(2)^2 + 13(2) - 10$

$= 16 + 52 + 26 - 10$

= 84

Therefore, $(x - 2)$ is not a factor.

If $x = -2$, $f(-2) = 2(-2)^3 + 13(-2)^2 + 13(-2) - 10$

$$= -16 + 52 - 26 - 10$$

$$= 0$$

Therefore, $(x + 2)$ is a factor of $f(x)$

We now divide $f(x)$ by $(x + 2)$ in order to get the other factors. This gives:

$$
\begin{array}{r}
2x^2 + 9x - 5 \\
x + 2 \overline{\smash{)}2x^3 + 13x^2 + 13x - 10} \\
\underline{2x^3 + 4x^2} \\
9x^2 + 13x \\
\underline{9x^2 + 18x} \\
-5x - 10 \\
\underline{-5x - 10} \\
- \quad - \quad -
\end{array}
$$

Hence, $\dfrac{2x^3 + 13x^2 + 13x - 10}{x + 2} = 2x^2 + 9x - 5$

Or, $2x^3 + 13x^2 + 13x - 10 = (x + 2)(2x^2 + 9x - 5)$

The quadratic part $2x^2 + 9x - 5$ can be factorized to give:

$2x^2 + 9x - 5 = 2x^2 + 10x - x - 5$

$$= 2x(x + 5) - 1\,(x + 5)$$

$$= (x + 5)(2x - 1)$$

$\therefore\quad 2x^3 + 13x^2 + 13x - 10 = (x + 2)(x + 5)(2x - 1)$

Let us now use the factorized expression above to solve the given equation as follows:

$2x^3 + 13x^2 + 13x - 10 = 0$

$(x + 2)(x + 5)(2x - 1) = 0$

Hence, $x = -2$ or -5 or $\dfrac{1}{2}$ (When each bracket is equated to zero and solved)

3. Find the remainder when $x^3 - 2x^2 + 5x + 9$ is divided by $x + 2$.

<u>Solution</u>

Let $f(x) = x^3 - 2x^2 + 5x + 9$

According to the remainder theorem, when this polynomial is divided by $x + 2$, the remainder is obtained from $f(-2)$. Note that -2 is obtained when we set $x + 2 = 0$ and solve it to get $x = -2$.

$\therefore\quad f(x) = x^3 - 2x^2 + 5x + 9$

$f(-2) = (-2)^3 - 2(-2)^2 + 5(-2) + 9$

$$= -8 - 8 - 10 + 9$$

$$= -17$$

Hence the remainder is −17

4. Find the remainder when $2x^2 + 8x - 3$ is divided by $3x - 1$

Solution

Let $f(x) = 2x^2 + 8x - 3$

According to the remainder theorem, when this polynomial is divided by $3x - 1$, the remainder is obtained from $f(\frac{1}{3})$. Note that $\frac{1}{3}$ is obtained when we set $3x - 1 = 0$ and solve it to get $x = \frac{1}{3}$.

$\therefore \quad f(x) = 2x^2 + 8x - 3$

$f(\frac{1}{3}) = 2(\frac{1}{3})^2 + 8(\frac{1}{3}) - 3$

$\quad = 2(\frac{1}{9}) + \frac{8}{3} - 3$

$\quad = \frac{2}{9} + \frac{8}{3} - 3$

$\quad = \frac{2 + 24 - 27}{9}$

$\quad = -\frac{1}{9}$

Hence the remainder is $-\frac{1}{9}$

5. If $(x + 1)$ and $(3x + 2)$ are factors of $3x^3 + 2x^2 - 3x - 2$, find the third factor.

Solution

Since $(x + 1)$ and $(3x + 2)$ are factors of $3x^3 + 2x^2 - 3x - 2$, then the third factor can be obtained by dividing $3x^3 + 2x^2 - 3x - 2$ by $(x + 1)(3x + 2)$. This is the simple logic:

$\quad (x + 1)(3x + 2)(\quad) = 3x^3 + 2x^2 - 3x - 2$ (Where ( ) is the third factor)

Hence, $(\quad) = \dfrac{3x^3 + 2x^2 - 3x - 2}{(x + 1)(3x + 2)}$ (When both sides of the equation are divide by $(x + 1)(3x + 2)$

$\quad (\quad) = \dfrac{3x^3 + 2x^2 - 3x - 2}{3x^2 + 5x + 2}$ (After expanding the denominator)

We now carry out the division as shown below

$$
\begin{array}{r}
x - 1 \\
3x^2 + 5x + 2 \overline{\smash{)}\ 3x^3 + 2x^2 - 3x - 2} \\
\underline{3x^3 + 5x^2 + 2x} \\
-3x^2 - 5x - 2 \\
\underline{-3x^2 - 5x - 2} \\
-\quad -\quad -\quad -
\end{array}
$$

Hence the third factor is $x - 1$

6. If $(3x - 1)$ is a factor of the polynomial $f(x) = 6x^2 + kx - 1$, where k is a constant, find the zeros of $f(x)$.

Solution

Since $3x - 1$ is a factor, then we obtain x as follows:

$3x - 1 = 0$

$3x = 1$

$x = \dfrac{1}{3}$

Hence, $f(\dfrac{1}{3}) = 0$

$f(x) = 6x^2 + kx - 1$

$f(\dfrac{1}{3}) = 6(\dfrac{1}{3})^2 + k(\dfrac{1}{3}) - 1$

$= 6(\dfrac{1}{9}) + \dfrac{k}{3} - 1$

$= \dfrac{6}{9} + \dfrac{k}{3} - 1$

$= \dfrac{2}{3} + \dfrac{k}{3} - 1$

$= \dfrac{2 + k - 3}{3}$

$= \dfrac{k - 1}{3}$

But $f(\dfrac{1}{3}) = 0$

Therefore, $\dfrac{k - 1}{3} = 0$

$k - 1 = 3(0)$

$k - 1 = 0$

$k = 1$

Hence the polynomial is $f(x) = 6x^2 + x - 1$ (Since k = 1)

In order to obtain the zeros of the polynomial, we solve the polynomial equation by factorization as follows:

$6x^2 + x - 1 = 0$

$6x^2 + 3x - 2x - 1 = 0$

$3x(2x + 1) - 1(2x + 1) = 0$

$(2x + 1)(3x - 1) = 0$

Therefore, $x = -\dfrac{1}{2}$ or $x = \dfrac{1}{3}$ (i.e. the roots of the equation)

Hence the zeros of $f(x)$ are $-\dfrac{1}{2}$ and $x = \dfrac{1}{3}$

7. Given that $x + 2$ is a factor of the polynomial $2x^3 - x^2 - 7x + 6$, find the other two factors.

Solution

Let us first divide $2x^2 - x^2 - 7x + 6$ by $x + 2$. This is as shown below.

$$2x^2 - 5x + 3$$
$$x + 2\overline{)2x^3 - x^2 - 7x + 6}$$
$$\underline{2x^3 + 4x^2}$$
$$-5x^2 - 7x$$
$$\underline{-5x^2 - 10x}$$
$$3x + 6$$
$$\underline{3x + 6}$$
$$-\quad-\quad-$$

Hence the quadratic factor of the polynomial is $2x^2 - 5x + 3$. Let us factorize this quadratic expression in order to find the other two linear factors as follows:

$$2x^2 - 5x + 3 = 2x^2 - 3x - 2x + 3$$
$$= x(2x - 3) - 1(2x - 3)$$
$$= (2x - 3)(x - 1)$$

Therefore the other two factors are $(2x - 3)$ and $(x - 1)$

8. The remainder when the polynomial $f(x) = ax^3 + bx^2 + x - 5$ is divided by $x + 2$ is -39, and when it is divided by $x - 1$ the remainder is -3. Determine the values of a and b.

Solution

$$f(x) = ax^3 + bx^2 + x - 5$$

When $f(x)$ is divided by $x + 2$, then $x = -2$ (When we set $x + 2 = 0$, then $x = -2$)

Hence, $f(-2) = a(-2)^3 + b(-2)^2 + (-2) - 5$

$$-39 = a(-8) + b(4) - 2 - 5 \quad\quad \text{(Note that } -39 \text{ is the remainder)}$$
$$-39 = -8a + 4b - 7$$
$$8a - 4b = 39 - 7$$
$$8a - 4b = 32$$
$$2a - b = 8 \quad\quad \text{(After dividing each term by 8)}$$
$$2a - b = 8 \;................\text{Equation (1)}$$

Similarly, when $f(x)$ is divided by $x - 1$, then $x = 1$ (When we set $x - 1 = 0$, then $x = 1$)

Hence, $f(1) = a(1)^3 + b(1)^2 + (1) - 5$

$$-3 = a(1) + b(1) + 1 - 5 \quad\quad \text{(Note that } -3 \text{ is the remainder)}$$
$$-3 = a + b - 4$$
$$4 - 3 = a + b$$
$$a + b = 1 \;..................\text{Equation (2)}$$

From equation (2), $a = 1 - b \;..................\text{Equation (3)}$

Substitute $1 - b$ for a in equation (1) as follows.

$$2a - b = 8 \;................\text{Equation (1)}$$
$$2(1 - b) - b = 8$$
$$2 - 2b - b = 8$$

$2 - 3b = 8$

$-3b = 8 - 2$

$-3b = 6$

$b = \dfrac{6}{-3}$

$b = -2$

Substitute -2 for b in equation (3)in order to find a.

$a = 1 - b$ ....Equation (3)

$= 1 - (-2)$

$= 1 + 2$

$a = 3$

Therefore, a = 3 and b = -2

9. The polynomial $x^3 + 2x^2 + mx + n$ is divisible by $x + 1$. It leaves a remainder of 12 when it is divided by $x - 2$.

a. Find m and n

b. Factorize the polynomial completely

c. Find the zeros of the polynomial

Solution

a. $x^3 + 2x^2 + mx + n$

Since the polynomial is divisible by $x + 1$, it means that $x + 1$ is a factor of the polynomial. Hence, $x = -1$ is a root of the polynomial equation. Hence substituting $x = -1$, gives zero as follows.

$x^3 + 2x^2 + mx + n$

$(-1)^3 + 2(-1)^2 + m(-1) + n = 0$

$-1 + 2 - m + n = 0$

$-m + n = -1$ ....Equation (1)

Also, since the polynomial leaves a remainder of 12 when divided by $(x - 2)$, it means that if we put $x = 2$ in the polynomial equation, we will obtain 12. This is done as follows.

$x^3 + 2x^2 + mx + n$

$(2)^3 + 2(2)^2 + m(2) + n = 0$

$8 + 8 + 2m + n = 12$

$2m + n = 12 - 8 - 8$

$2m + n = -4$ ....Equation (2)

From equation (1), n = m − 1 ....Equation (3)

Substitute m − 1 for n in equation (2)

$2m + n = -4$ ....Equation (2)

$2m + (m - 1) = -4$

$2m + m - 1 = -4$

$3m = -4 + 1$

$$3m = -3$$
$$m = \frac{-3}{3}$$
$$m = -1$$

Substitute −1 for m in equation (3)

$$n = m - 1 \quad \ldots\ldots\ldots\ldots\ldots\text{Equation (3)}$$
$$= -1 - 1$$
$$n = -2$$

Hence, m = −1 and n = −2

b. The polynomial is $x^3 + 2x^2 + mx + n$. When we substitute the values of m and n the polynomial becomes:

$$x^3 + 2x^2 - x - 2 \qquad \text{(Since m = −1 and n = −2)}$$

From the question, $(x + 1)$ is a factor of the polynomial. Hence, let us divide the polynomial by $(x + 1)$ in order to get the other factors. The working is as shown below.

$$
\begin{array}{r}
x^2 + x - 2 \\
x + 1\overline{)x^3 + 2x^2 - x - 2} \\
\underline{x^3 + x^2} \\
x^2 - x \\
\underline{x^2 + x} \\
-2x - 2 \\
\underline{-2x - 2} \\
-\ -\ -
\end{array}
$$

Hence the quadratic factor is $x^2 + x - 2$. We now factorize it as follows:

$$x^2 + x - 2 = (x + 2)(x - 1)$$

Let us now write the factorized polynomial as follows:

$$x^3 + 2x^2 - x - 2 = (x + 1)(x + 2)(x - 1)$$

c. Set the factorized polynomial to be equal to zero. This gives:

$$x^3 + 2x^2 - x - 2 = 0$$
$$(x + 1)(x + 2)(x - 1) = 0$$

Hence, $x = -1, -2$ or 1 \qquad (When each of the bracket above is equated to zero and solved)

Therefore the zeros of the polynomial are −1, −2 and 1.

10. If $2x^2 - 7x - 4$ is a factor of the polynomial $f(x) = 2x^4 - 5x^3 - 15x^2 + px + q$, where p and q are constants:

a. find the values of p and q

b. factorize $f(x)$ completely.

Solution

a. Let us factorize $2x^2 - 7x - 4$ as follows:

$2x^2 - 7x - 4 = 2x^2 - 8x + x - 4$

$= 2x(x - 4) + 1(x - 4)$

$= (x - 4)(2x + 1)$

Hence $2x^2 - 7x - 4 = (x - 4)(2x + 1)$

Therefore $(x - 4)$ and $(2x + 1)$ are factors of f(x)

From these two factors, $x = 4$ and $x = -\dfrac{1}{2}$

Hence, f(4) = 0 and f$(-\dfrac{1}{2})$ = 0

$\quad$ f(x) = $2x^4 - 5x^3 - 15x^2 + px + q$

$\quad$ f(4) = $2(4)^4 - 5(4)^3 - 15(4)^2 + p(4) + q$

Since f(4) = 0, then it follows that:

$\quad 2(4)^4 - 5(4)^3 - 15(4)^2 + p(4) + q = 0$

$\quad 2(256) - 5(64) - 15(16) + 4p + q = 0$

$\quad 512 - 320 - 240 + 4p + q = 0$

$\quad - 48 + 4p + q = 0$

$\quad 4p + q = 48$ ....Equation (1)

Also, f(x) = $2x^4 - 5x^3 - 15x^2 + px + q$

$\quad$ f$(-\dfrac{1}{2})$ = $2(-\dfrac{1}{2})^4 - 5(-\dfrac{1}{2})^3 - 15(-\dfrac{1}{2})^2 + p(-\dfrac{1}{2}) + q$

Since f$(-\dfrac{1}{2})$ = 0, then it follows that:

$\quad 2(-\dfrac{1}{2})^4 - 5(-\dfrac{1}{2})^3 - 15(-\dfrac{1}{2})^2 + p(-\dfrac{1}{2}) + q = 0$

$\quad 2(\dfrac{1}{16}) - 5(-\dfrac{1}{8}) - 15(\dfrac{1}{4}) - \dfrac{p}{2} + q = 0$

$\quad \dfrac{1}{8} + \dfrac{5}{8} - \dfrac{15}{4} - \dfrac{p}{2} + q = 0$

$\quad \dfrac{1 + 5 - 30 - 4p + 8q}{8} = 0$

$\quad - 24 - 4p + 8q = 8(0)$

$\quad - 24 = 4p - 8q$

$\quad 4p - 8q = - 24$

$\quad p - 2q = - 6 \qquad$ (After dividing each term by 4)

$\quad p - 2q = - 6$ ....Equation (2)

From equation (2), p = 2q - 6 ....Equation (3)

Substitute 2q - 6 for p in equation 1.

$\quad 4p + q = 48$ ....Equation (1)

$\quad 4(2q - 6) + q = 48$

$\quad 8q - 24 + q = 48$

$9q = 48 + 24$

$9q = 72$

$q = \dfrac{72}{9}$

$q = 8$

Substitute 8 for q in equation (3)

$p = 2q - 6$ ....Equation (3)

$= 2(8) - 6$

$= 16 - 6$

$p = 10$

Hence, p = 10 and q = 8

b. Substitute 10 for p and 8 for q in order to obtain the polynomial as follows:

$f(x) = 2x^4 - 5x^3 - 15x^2 + 10x + 8$

Since $2x^2 - 7x - 4$ is a factor of the polynomial, let us divide f(x) by $2x^2 - 7x - 4$ as shown below.

$$
\begin{array}{r}
x^2 + x - 2 \\
2x^2 - 7x - 4{\overline{\smash{\big)}\,2x^4 - 5x^3 - 15x^2 - 10x + 8}} \\
\underline{2x^4 - 7x^3 - 4x^2} \\
2x^3 - 11x^2 + 10x \\
\underline{2x^3 - 7x^2 - 4x} \\
-4x^2 + 14x + 8 \\
\underline{-4x^2 + 14x + 8} \\
- - - - - -
\end{array}
$$

Hence the other quadratic factor is $x^2 + x - 2$

We now factorize this as follows:

$x^2 + x - 2 = (x + 2)(x - 1)$

Recall that: $2x^2 - 7x - 4 = (x - 4)(2x + 1)$ (From our first step in (a) above)

Therefore the polynomial is completely factorized by using all our linear factors as follows:

$2x^4 - 5x^3 - 15x^2 + 10x + 8 = (x + 2)(x - 1)(x - 4)(2x + 1)$

Exercise 9

1. If $A = 3x^3 + 5x^2 - 1$, $B = 2x^3 + 7$, and $C = 2x^3 - x^2 + 2x - 4$, find:

a. $A + B$

b. $2C - 2B$

c. $3C - A$

d. $2B + A - 2C$

e. $A + 2B + C$

2. If $f(x) = x^3 - 4x^2 + 3x - 10$, find:

a. $f(-3)$

b. $f(0)$

3. If $A = 5x^3 - 2x^2 + x$ and $B = 2x^3 + 3x^2 - 5x - 2$, find AB.

4. If $M = 2x^3 - 3x^2 + 6x + 5$ and $N = 3x^3 - x^2 + 2x - 4$, find MN.

5. Given that $f(x) = x^3 - 5$ and $g(x) = 2x^3 - 3x^2 + 6x - 9$, find $f(x) \cdot g(x)$

6. Divide $3x^2 + 2x - 8$ by $3x - 4$

7. Divide $2x^3 + x^2 - 7x + 24$ by $x + 3$

8. Evaluate $\dfrac{6x^3 + 3x^2 - 10x - 5}{2x + 1}$

9. Divide $4x^3 - 12x^2y + 11xy^2 + 15y^3$ by $2x - 5y$

10. Simplify: $\dfrac{5x^3 + 8x^2y - 5xy^2 - 2y^3}{x + 2y}$

11. Find the quotient and the remainder when $2x^3 - 7x^2 + 12x - 9$ is divided by $x - 5$

12. Find the quotient and the remainder when $5x^3 + 2x^2 - 11x + 1$ is divided by $x + 2$

13. Find the zeros of the following polynomial functions:

a. $f(x) = 2x^2 - 19x + 30$

b. $f(x) = 4x^2 + 4x - 3$

14. Factorize $x^3 - 4x^2 + x + 6$ and use it to solve the cubic equation $x^3 - 4x^2 + x + 6 = 0$

15. Solve the equation $2x^4 + x^3 - 20x^2 - 13x + 30 = 0$

16. Find the remainder when $2x^3 - 9x^2 + 6x + 3$ is divided by $x - 4$.

17. Find the remainder when $5x^2 - 2x - 7$ is divided by $2x + 3$

18. If $(x + 1)$ and $(x - 2)$ are factors of $4x^4 + 4x^3 - 13x^2 - 19x - 6$, find the other two factors.

19. If $(2x - 1)$ is a factor of the polynomial $f(x) = 12x^3 + kx$, where k is a constant, find the zeros of $f(x)$.

20. Given that $2x + 1$ is a factor of the polynomial $6x^3 + 3x^2 - 10x - 5$, find the quadratic factor.

21. The remainder when the polynomial $f(x) = ax^3 + bx^2 + 5x - 9$ is divided by $x - 1$ is -10, and when it is divided by $x - 3$ the remainder is -12. Determine the values of a and b.

22. The polynomial $x^3 - 2x^2 + mx + n$ is divisible by $x - 1$. It leaves a remainder of -8 when divided by $x - 2$.

a. Find m and n

b. Find the quadratic factor

c. Find the zeros of the polynomial

23. If $2x^2 - x - 6$ is a factor of the polynomial $f(x) = 2x^4 + x^3 - 20x^2 + mx + n$, where m and n are constants. Find the values of m and n

CHAPTER 10
PARTIAL FRACTION

Consider the addition of the algebraic fractions below.

$$\frac{3}{x-2} + \frac{2}{x+5} = \frac{3(x+5) + 2(x-2)}{(x-2)(x+5)}$$ [Note that the LCM is $(x-2)(x+5)$]

$$= \frac{3x+15+2x-4}{(x-2)(x+5)}$$

$$= \frac{5x+11}{(x-2)(x+5)}$$

Hence, $\frac{3}{x-2}$ and $\frac{2}{x+5}$ added up to give $\frac{5x+11}{(x-2)(x+5)}$

We say that $\frac{3}{x-2}$ and $\frac{2}{x+5}$ are the partial fractions of $\frac{5x+11}{(x-2)(x+5)}$. The result obtained i.e.

$\frac{5x+11}{(x-2)(x+5)}$ is called the compound fraction. Since we know how to move from partial

fractions to compound fraction, it is also important that we know how to move from compound fraction to partial fractions. This is what we want to deal with here.

Resolving Algebraic Fractions into Partial Fractions

The steps below are applied in resolving algebraic fraction into partial fractions:

1. The polynomial must be factorized or already be expressed in factors.

2. The highest power of the variable in the numerator must be at least one less than that of the denominator.

3. When the highest power of the variable in the numerator is equal to, or higher than the highest power of the variable in the denominator, then the numerator should be divided by the denominator before resolving the remainder (expressed as fraction) into partial fractions.

Types of Partial Fraction

1. **LINEAR FACTORS:** Algebraic fractions whose denominators contain linear factors are resolved into partial fractions as shown below.

$$\frac{f(x)}{(x+a)(x+b)} = \frac{A}{x+a} + \frac{B}{x+b}$$

2. **REPEATED LINEAR FACTORS:** Fractions with repeated linear denominators are resolved as follows:

$$\frac{f(x)}{(x+a)^2} = \frac{A}{x+a} + \frac{B}{(x+a)^2}$$

And:

$$\frac{f(x)}{(x+a)^3} = \frac{A}{x+a} + \frac{B}{(x+a)^2} + \frac{C}{(x+a)^3}$$

3. **QUADRATIC FACTORS:** When a denominator of an algebraic fraction contains a quadratic expression that cannot be factorized into linear factors, then it is resolved into partial fractions as follows:

$$\frac{f(x)}{(ax^2+bx+c)(x+d)} = \frac{Ax+B}{(ax^2+bx+c)} + \frac{C}{x+d}$$

Note that in all three cases above, the highest power of the variable in f(x) must be less than the highest power of the variable in the denominator. Otherwise we must divide first before we proceed.

Examples

1. Express $\dfrac{x+21}{(x+3)(x+1)}$ in partial fractions

<u>Solution</u>

METHOD 1

Let $\dfrac{x+21}{(x+3)(x+1)} = \dfrac{A}{x+3} + \dfrac{B}{x+1}$

The LCM or LCD (lowest common denominator) of the right hand side is $(x+3)(x+1)$. Hence we carry out the addition of the right hand side as follows:

$$\frac{x+21}{(x+3)(x+1)} = \frac{A(x+1)+B(x+3)}{(x+3)(x+1)}$$

Hence, since the denominators are the same on both sides of the equation, then the numerators should also be the same or identical. Hence for the numerators, we write:

$x + 21 \equiv A(x+1) + B(x+3)$

This is an identity. It means that both sides are equal for all values of x, hence the use of the equivalent symbol. In order to find A and B, we take a value of x that will make the term in any of the bracket to be zero. In order to find the number that will make A to be zero, we simply equate the bracket attached to A to be equal to zero and solve for x. For example, $x + 1 = 0$, which will give $x = -1$ when we solve for x. Hence if we put $x = -1$, A will be eliminated so that we can obtain B. This gives:

$x + 21 \equiv A(x + 1) + B(x + 3)$

$-1 + 21 = A(-1 + 1) + B(-1 + 3)$

$20 = A(0) + 2B$

$20 = 2B$ (Since A x 0 = 0)

$B = \dfrac{20}{2}$

$B = 10$

If we put $x = -3$ (from $x + 3 = 0$, then $x = -3$), B will be eliminated, so that we can obtain A. This gives:

$x + 21 \equiv A(x + 1) + B(x + 3)$

$-3 + 21 = A(-3 + 1) + B(-3 + 3)$

$$18 = A(-2) + B(0)$$

$$18 = -2A \quad (\text{Since B x 0 = 0})$$

$$A = \frac{18}{-2}$$

$$A = -9$$

Hence, A = −9 and B = 10

Therefore, $\dfrac{x + 21}{(x + 3)(x + 1)} = \dfrac{-9}{x + 3} + \dfrac{10}{x + 1}$

Or, $\dfrac{x + 21}{(x + 3)(x + 1)} = \dfrac{10}{x + 1} - \dfrac{9}{x + 3}$ (When we rearrange the answer above)

METHOD 2

In this method, we follow the same procedure as in method 1 until we get to the point written below:

$x + 21 \equiv A(x + 1) + B(x + 3)$

Now we expand the brackets above, then simplify and collect like terms as carried out below.

$x + 21 \equiv Ax + A + Bx + 3B$

$x + 21 \equiv Ax + Bx + A + 3B$ (By collecting like terms together)

$x + 21 \equiv (A + B)x + A + 3B$ (After factorizing terms having like variables)

By comparing coefficients of like terms on both sides of the identity, we have:

$A + B = 1$

On the left hand side of the identity above, we have $1x$, while on the right hand side we have $(A + B)x$. Hence the coefficients of x are 1 and A + B. Therefore, A + B = 1.

Similarly, $A + 3B = 21$

The constant term on the left hand side is 21, while the constant terms on the right hand side are A and 3B. Hence we equate them to obtain A + 3B = 21.

Therefore our two equations above are:

$$A + B = 1 \text{................Equation (1)}$$

$$\underline{A + 3B = 21} \text{.................Equation (2)}$$

Equation (2) − equation (1): $2B = 20$

$$B = \frac{20}{2}$$

$$B = 10$$

Substitute 10 for B in equation (1).

$A + B = 1$ ....Equation (1)

$A + 10 = 1$

$A = 1 - 10$

$A = -9$

Therefore, A = −9 and B = 10

Hence, $\dfrac{x + 21}{(x + 3)(x + 1)} = \dfrac{-9}{x + 3} + \dfrac{10}{x + 1}$

Or, $\dfrac{x + 21}{(x + 3)(x + 1)} = \dfrac{10}{x + 1} - \dfrac{9}{x + 3}$ (As obtained in method 1 above)

It is obvious that method 1 is more direct or shorter than method 2, since method 2 leads to a simultaneous equation. Hence, in most of our examples we will be using method 1. However in some cases where only method 1 cannot be used, we will apply the two methods or only method 2, as the case may be.

2.   Resolve $\dfrac{4x - 16}{x^2 - 2x - 3}$ into partial fractions.

Solution

The denominator can be factorized into linear factors as follows:

$x^2 - 2x - 3 = (x - 3)(x + 1)$

Hence the fraction is now written as follows:

$\dfrac{4x - 16}{x^2 - 2x - 3} = \dfrac{4x - 16}{(x - 3)(x + 1)}$

We now resolve the factorized one into partial fractions as follows:

Let $\dfrac{4x - 16}{(x - 3)(x + 1)} = \dfrac{A}{x - 3} + \dfrac{B}{x + 1}$

$\dfrac{4x - 16}{(x - 3)(x + 1)} = \dfrac{A(x + 1) + B(x - 3)}{(x - 3)(x + 1)}$

Since the denominators are the same, then the numerators are identical and can be brought out to give:

$4x - 16 \equiv A(x + 1) + B(x - 3)$

Now substitute $x = -1$ into the identity in order to obtain B as follows:

$4x - 16 \equiv A(x + 1) + B(x - 3)$

$4(-1) - 16 = A(-1 + 1) + B(-1 - 3)$

$-4 - 16 = A(0) - 4B$

$-20 = -4B$ (Since A x 0 = 0)

$B = \dfrac{-20}{-4}$

$B = 5$

Put $x = 3$ in order to obtain A as follows:

$4x - 16 \equiv A(x + 1) + B(x - 3)$

$4(3) - 16 = A(3 + 1) + B(3 - 3)$

$12 - 16 = A(4) + B(0)$

$-4 = 4A$ (Since B x 0 = 0)

$A = \dfrac{-4}{4}$

$A = -1$

133

Hence, A = −1 and B = 5

Therefore, $\dfrac{4x - 16}{x^2 - 2x - 3} = -\dfrac{1}{x - 3} + \dfrac{5}{x + 1}$

$\qquad\qquad\quad = \dfrac{5}{x + 1} - \dfrac{1}{x - 3}$ (When we rearrange the answer above)

3. Express $\dfrac{15}{x^2 - 9}$ in partial fractions.

<u>Solution</u>

The denominator is a difference of two squares. Recall that a difference of two squares such as $a^2 - b^2$ can be factorized as follows:

$\qquad a^2 - b^2 = (a + b)(a - b)$

Similarly, the denominator above can be expressed as a difference of two squares and factorized as follows:

$\qquad x^2 - b^2 = x^2 - 3^2 = (x + 3)(x - 3)$ (Note that 9 has been expressed as 3^2)

Hence, $\dfrac{15}{(x + 3)(x - 3)} = \dfrac{A}{x + 3} + \dfrac{B}{x - 3}$

$\dfrac{15}{(x + 3)(x - 3)} = \dfrac{A(x - 3) + B(x + 3)}{(x + 3)(x - 3)}$

$15 \equiv A(x - 3) + B(x + 3)$

Substitute $x = 3$ into the identity to obtain B as follows:

$\quad 15 = A(3 - 3) + B(3 + 3)$ (Note that the right hand side remains 15 since there is no x there)

$\quad 15 = 0 + 6B$

$\quad 15 = 6B$

$\quad B = \dfrac{15}{6}$

$\quad B = \dfrac{5}{2}$ (In its lowest term)

Put $x = -3$ in order to eliminate B and obtain A as follows:

$\quad 15 \equiv A(x - 3) + B(x + 3)$

$\quad 15 = A(-3 - 3) + B(-3 + 3)$

$\quad 15 = -6A + 0$

$\quad 15 = -6A$

$\quad A = \dfrac{15}{-6}$

$\quad A = -\dfrac{5}{2}$

Hence, $A = -\dfrac{5}{2}$ and $B = \dfrac{5}{2}$

Therefore, $\dfrac{15}{x^2 - 9} = \dfrac{-\dfrac{5}{2}}{x + 3} + \dfrac{\dfrac{5}{2}}{x - 3}$

$$= -\frac{5}{2(x+3)} + \frac{5}{2(x-3)}$$

$$= \frac{5}{2(x-3)} - \frac{5}{2(x+3)}$$

4. Resolve $\dfrac{2x-9}{x(x-3)}$ into partial fractions.

Solution

$$\frac{2x-9}{x(x-3)} = \frac{A}{x} + \frac{B}{x-3}$$

$$\frac{2x-9}{x(x-3)} = \frac{A(x-3) + B(x)}{x(x-3)}$$

Equating the numerators gives:

$$2x - 9 \equiv A(x-3) + B(x)$$

Substitute $x = 3$ into the identity to obtain B as follows:

$$2(3) - 9 = A(3-3) + B(3)$$

$$6 - 9 = 0 + 3B$$

$$-3 = 3B$$

$$B = \frac{-3}{3}$$

$$B = -1$$

Put $x = 0$ in order to eliminate B and obtain A as follows:

$$2x - 9 \equiv A(x-3) + B(x)$$

$$2(0) - 9 = A(0-3) + B(0)$$

$$0 - 9 = -3A + 0$$

$$-9 = -3A$$

$$A = \frac{-9}{-3}$$

$$A = 3$$

Hence, A = 3 and B = −1

Therefore, $\dfrac{2x-9}{x(x-3)} = \dfrac{3}{x} - \dfrac{1}{x-3}$

5. Resolve $\dfrac{2x+12}{(x+5)^2}$ into partial fractions.

Solution

The denominator is a repeated linear factor. Hence we resolve into partial factions as follows:

$$\frac{2x+12}{(x+5)^2} = \frac{A}{x+5} + \frac{B}{(x+5)^2}$$

$$\frac{2x+12}{(x+5)^2} = \frac{A(x+5) + B(1)}{(x+5)^2} \qquad \text{[Note that } (x+5)^2 = (x+5)(x+5)\text{]}$$

Equating the numerators gives:

$$2x + 12 \equiv A(x+5) + B$$

Substitute –5 for x in order to solve for B

$2(-5) + 12 = A(-5 + 5) + B$

$-10 + 12 = A(0) + B$

$2 = B$

$B = 2$

Since we already know the value of B, we can substitute any value for x into the identity above and obtain the value of A. Note that an identity is true for all values of the variable. Hence let us substitute 1 for x (any value of x can be used) into the identity while also substituting 2 for B as we have already obtained above. This will give us A as follows:

$2x + 12 \equiv A(x + 5) + B$

$2(1) + 12 = A(1 + 5) + 2$

$2 + 12 = 6A + 2$

$14 - 2 = 6A$

$12 = 6A$

$A = \dfrac{12}{6}$

$A = 2$

Hence, A = 2 and B = 2

Therefore, $\dfrac{2x + 12}{(x + 5)^2} = \dfrac{2}{x + 5} + \dfrac{2}{(x + 5)^2}$

6. Express $\dfrac{3x - 5}{x^2(x-2)}$ in partial fractions.

Solution

x^2 in the denominator is like a repeated linear factor. With that in mind, we express in partial fraction as follows:

$\dfrac{3x - 5}{x^2(x - 2)} = \dfrac{A}{x} + \dfrac{B}{x^2} + \dfrac{C}{x - 2}$

$\dfrac{3x - 5}{x^2(x - 2)} = \dfrac{A[x(x - 2)] + B(x - 2) + C(x^2)}{x^2(x - 2)}$

Equating the numerators gives:

$3x - 5 \equiv A[x(x - 2)] + B(x - 2) + Cx^2$

Observing the right hand side of the identity shows that if we substitute $x = 0$, we will obtain the value of B as follows:

$3x - 5 \equiv A[x(x - 2)] + B(x - 2) + Cx^2$

$3(0) - 5 = A[0(0 - 2)] + B(0 - 2) + C(0^2)$

$0 - 5 = A[0(-2)] + B(-2) + 0$

$-5 = 0 - 2B + 0$

$-5 = -2B$

$$B = \frac{-5}{-2}$$

$$B = \frac{5}{2}$$

Similarly, substituting 2 for x in the identity above will give us the value of C as follows:

$$3x - 5 \equiv A[x(x - 2)] + B(x - 2) + Cx^2$$

$$3(2) - 5 = A[2(2 - 2)] + B(2 - 2) + C(2^2)$$

$$6 - 5 = A[2(0)] + B(0) + 4C$$

$$1 = 0 + 0 + 4C$$

$$1 = 4C$$

$$C = \frac{1}{4}$$

Now, since we already know the values of B and C, we can substitute any value of x (apart from 0 and 2 used above) into the identity in order to obtain the last letter which is A. Note that this can be done when finding the last letter in any identity. Hence, let us put $x = 1$ into the identity as follows:

$$3x - 5 \equiv A[x(x - 2)] + B(x - 2) + Cx^2$$

$$3(1) - 5 = A[1(1 - 2)] + \frac{5}{2}(1 - 2) + \frac{1}{4}(1^2)$$

$$3 - 5 = A[1(-1)] + \frac{5}{2}(-1) + \frac{1}{4}(1)$$

$$-2 = -A - \frac{5}{2} + \frac{1}{4}$$

$$A = 2 - \frac{5}{2} + \frac{1}{4}$$

$$A = \frac{8 - 10 + 1}{4}$$

$$A = -\frac{1}{4}$$

Hence, $A = -\frac{1}{4}$, $B = \frac{5}{2}$ and $C = \frac{1}{4}$

Therefore, $\dfrac{3x - 5}{x^2(x-2)} = \dfrac{-\frac{1}{4}}{x} + \dfrac{\frac{5}{2}}{x^2} + \dfrac{\frac{1}{4}}{x - 2}$

$$= -\frac{1}{4x} + \frac{5}{2x^2} + \frac{1}{4(x - 2)}$$

This can also be rearranged to avoid starting with a negative value as follows:

$$\frac{3x - 5}{x^2(x-2)} = \frac{5}{2x^2} - \frac{1}{4x} + \frac{1}{4(x - 2)}$$

7. Resolve $\dfrac{x + 6}{(x + 2)(x - 3)^2}$ into partial fractions.

Solutions

The denominator contains repeated factors. Hence we resolve as follows:

137

$$\frac{x+6}{(x+2)(x-3)^2} = \frac{A}{x+2} + \frac{B}{x-3} + \frac{C}{(x-3)^2}$$

$$\frac{x+6}{(x+2)(x-3)^2} = \frac{A(x-3)^2 + B(x+2)(x-3) + C(x+2)}{(x+2)(x-3)^2}$$

Equating the numerators gives:

$$x + 6 \equiv A(x-3)^2 + B(x+2)(x-3) + C(x+2)$$

Substitute 3 for x in order to obtain C as follows:

$$3 + 6 = A(3-3)^2 + B(3+2)(3-3) + C(3+2)$$

$$9 = A(0) + B(5)(0) + C(5)$$

$$9 = 0 + 0 + 5C$$

$$9 = 5C$$

$$C = \frac{9}{5}$$

Substitute -2 for x in the identity above in order to find A. This gives:

$$x + 6 \equiv A(x-3)^2 + B(x+2)(x-3) + C(x+2)$$

$$-2 + 6 = A(-2-3)^2 + B(-2+2)(-2-3) + C(-2+2)$$

$$4 = A(-5)^2 + B(0)(-5) + C(0)$$

$$4 = A(25) + 0 + 0$$

$$4 = 25A$$

$$A = \frac{4}{25}$$

Since we now know the values of A and C, we can substitute any value for x in the identity above in order to get B. However, I want us to use the method of comparing coefficient to find B. In order to use this method, we have to expand the brackets on the right hand side of the identity. Let us expand and then compare coefficients as follows:

$$x + 6 \equiv A(x-3)^2 + B(x+2)(x-3) + C(x+2)$$

$$x + 6 \equiv A(x-3)(x-3) + B(x+2)(x-3) + C(x+2)$$

$$x + 6 \equiv A(x^2 - 6x + 9) + B(x^2 - x - 6) + Cx + 2C$$

$$x + 6 \equiv Ax^2 - 6Ax + 9A + Bx^2 - Bx - 6B + Cx + 2C$$

$$x + 6 \equiv (A + B)x^2 - (6A + B - C)x + (9A - 6B + 2C)$$

By comparing the coefficient of x^2 on both sides of the identity shows that:

$$A + B = 0$$

Since there is no term in x^2 on the left hand side of the identity, it means that the coefficient of x^2 on the left hand side is zero. The coefficient of x^2 on the right hand side is A + B. Hence equating the two coefficients gives A + B = 0, as written above.

We already know the value of A from our answer above, hence we calculate B as follows:

$$A + B = 0$$

$$\frac{4}{25} + B = 0$$

$$B = -\frac{4}{25}$$

Hence, $A = \frac{4}{25}$, $B = -\frac{4}{25}$ and $C = \frac{9}{5}$

Therefore, $\dfrac{x+6}{(x+2)(x-3)^2} = \dfrac{4}{25(x+2)} - \dfrac{4}{25(x-3)} + \dfrac{9}{5(x-3)^2}$

Or, $\dfrac{x+6}{(x+2)(x-3)^2} = \dfrac{1}{5}\left[\dfrac{4}{5(x+2)} - \dfrac{4}{5(x-3)} + \dfrac{9}{(x-3)^2}\right]$

8. Resolve $\dfrac{2x^2-3x-4}{x(x^2-2)}$ into partial fractions.

Solution

Note that the denominator contains a quadratic factor which does not factorize. This means that when the quadratic factor is resolved to partial fraction, its numerator will contain x and two constant terms. Hence we resolve into partial fractions as follows:

$$\frac{2x^2-3x-4}{x(x^2-2)} = \frac{A}{x} + \frac{Bx+C}{x^2-2}$$

$$\frac{2x^2-3x-4}{x(x^2-2)} = \frac{A(x^2-2)+(Bx+C)(x)}{x(x^2-2)}$$

Equating the numerators gives:

$$2x^2 - 3x - 4 \equiv A(x^2 - 2) + (Bx + C)(x)$$

Substitute 0 for x in the identity above. This gives:

$$2(0)^2 - 3(0) - 4 = A((0)^2 - 2) + [B(0) + C](0)$$

$$0 - 0 - 4 = A(0 - 2) + 0$$

$$-4 = -2A$$

$$A = \frac{-4}{-2}$$

$$A = 2$$

In the identity above, we can no longer substitute any value of x in order to obtain B or C. Hence we have to use the method of comparing coefficients to find B and C. In order to use this method, we expand the bracket in the identity above. This gives:

$$\begin{aligned}2x^2 - 3x - 4 &\equiv A(x^2 - 2) + (Bx + C)(x) \\ &\equiv Ax^2 - 2A + Bx^2 + Cx \\ &\equiv Ax^2 + Bx^2 + Cx - 2A\end{aligned}$$

$$2x^2 - 3x - 4 \equiv (A + B)x^2 + Cx - 2A \quad \text{[Note that } Ax^2 + Bx^2 \text{ has been factorized to give } (A + B)x^2\text{]}$$

Comparing the coefficients of x^2 on both sides of the identity shows that A + B = 2. This is because on the left hand side we have $2x^2$ and on the right hand side we have $(A + B)x^2$. Hence $2x^2 = (A + B)x^2$, which shows that A + B = 2.

Let us solve this equation as follows:

A + B = 2

2 + B = 2   (Since A = 2 as obtained above)

B = 2 − 2

139

B = 0

Similarly, comparing the coefficients of x in the identity above, shows that:

C = –3   (Since $-3x = Cx$)

Hence, A = 2, B = 0 and C = –3

Therefore, $\dfrac{2x^2 - 3x - 4}{x(x^2 - 2)} = \dfrac{2}{x} + \dfrac{0x - 3}{x^2 - 2}$

This gives: $\dfrac{2x^2 - 3x - 4}{x(x^2 - 2)} = \dfrac{2}{x} - \dfrac{3}{x^2 - 2}$

9. Resolve $\dfrac{2x^2 - 1}{(x^2 - 1)(x - 2)}$ into partial fractions

Solution

$$\frac{2x^2 - 1}{(x^2 - 1)(x - 2)} = \frac{Ax + B}{(x^2 - 1)} + \frac{C}{(x - 2)}$$

$$\frac{2x^2 - 1}{(x^2 - 1)(x - 2)} = \frac{(Ax + B)(x - 2) + C(x^2 - 1)}{(x^2 - 1)(x - 2)}$$

Equating the numerators gives:

$$2x^2 - 1 \equiv (Ax + B)(x - 2) + C(x^2 - 1)$$

Substitute 2 for x in order to get C. This gives:

$2(2)^2 - 1 = (A(2) + B)(2 - 2) + C(2^2 - 1)$

$2(4) - 1 = (2A + B)(0) + C(4 - 1)$

$8 - 1 = 0 + 3C$

$7 = 3C$

$C = \dfrac{7}{3}$

Now let us expand the bracket in the identity above.

$2x^2 - 1 \equiv (Ax + B)(x - 2) + C(x^2 - 1)$

$\equiv Ax^2 - 2Ax + Bx - 2B + Cx^2 - C$

$\equiv Ax^2 + Cx^2 - 2Ax + Bx - 2B - C$

$2x^2 - 1 \equiv (A + C)x^2 + (-2A + B)x - (2B + C)$ (After factorization)

Comparing coefficient of terms on both sides of the identity shows that:

A + C = 2 ………………..Equation (1)

–2A + B = 0 ………………Equation (2)

Note that there is no term in x on the left hand side, hence the coefficient of x on the left hand side is zero. This is why equation (2) is equated to zero as shown above.

–(2B + C) = –1   (This is obtained by comparing the coefficients of the constant terms)

Or, 2B + C = 1 ………………Equation (3)   (After dividing both sides by –1)

Hence from equation (1), we have:

A + C = 2 ………………..Equation (1)

140

$A + \dfrac{7}{3} = 2$ (Since $C = \dfrac{7}{3}$)

$A = 2 - \dfrac{7}{3}$

$= \dfrac{6 - 7}{3}$

$A = -\dfrac{1}{3}$

From equation (2) we have:

$-2A + B = 0$ ....Equation (2)

$-2\left(-\dfrac{1}{3}\right) + B = 0$

$\dfrac{2}{3} + B = 0$

$B = -\dfrac{2}{3}$

Hence, $A = -\dfrac{1}{3}$, $B = -\dfrac{2}{3}$ and $C = \dfrac{7}{3}$

Therefore, $\dfrac{2x^2 - 1}{(x^2 - 1)(x - 2)} = \dfrac{-\dfrac{1}{3}x - \dfrac{2}{3}}{(x^2 - 1)} + \dfrac{\dfrac{7}{3}}{(x - 2)}$

$= \dfrac{-\dfrac{1}{3}(x + 2)}{(x^2 - 1)} + \dfrac{7}{3(x - 2)}$

$\dfrac{2x^2 - 1}{(x^2 - 1)(x - 2)} = \dfrac{7}{3(x - 2)} - \dfrac{(x + 2)}{3(x^2 - 1)}$

Note that the method of comparing coefficients is often used when we have a quadratic factor in the denominator.

10. Resolve $\dfrac{6x^2 - 24x - 3}{(2x - 1)(x^2 + 5x + 4)}$ into partial fractions

Solution

A careful look at the quadratic part of the denominator shows that it can be factorized into linear factors as follows:

$x^2 + 5x + 4 = (x + 4)(x + 1)$

Hence the fraction in the question above can be rewritten as:

$\dfrac{6x^2 - 24x - 3}{(2x - 1)(x + 4)(x + 1)}$

We now resolve into partial fractions as follows:

$\dfrac{6x^2 - 24x - 3}{(2x - 1)(x + 4)(x + 1)} = \dfrac{A}{2x - 1} + \dfrac{B}{x + 4} + \dfrac{C}{x + 1}$

$\dfrac{6x^2 - 24x - 3}{(2x - 1)(x + 4)(x + 1)} = \dfrac{A(x + 4)(x + 1) + B(2x - 1)(x + 1) + C(2x - 1)(x + 4)}{(2x - 1)(x + 4)(x + 1)}$

Equating the numerators gives:

$6x^2 - 24x - 3 \equiv A(x + 4)(x + 1) + B(2x - 1)(x + 1) + C(2x - 1)(x + 4)$

Substitute $x = -4$ in order to obtain B. This gives:

$6(-4)^2 - 24(-4) - 3 = A(-4 + 4)(-4 + 1) + B(2(-4) - 1)(-4 + 1) + C(2(-4) - 1)(-4 + 4)$

$6(16) + 96 - 3 = A(0)(-3) + B(-8 - 1)(-3) + C(-8 - 1)(0)$

$96 + 96 - 3 = 0 + 27B + 0$

$189 = 27B$

$B = \dfrac{189}{27}$

$B = 7$

Substitute −1 for x in the identity above. This will give C as follows:

$6x^2 - 24x - 3 \equiv A(x + 4)(x + 1) + B(2x - 1)(x + 1) + C(2x - 1)(x + 4)$

$6(-1)^2 - 24(-1) - 3 = A(-1 + 4)(-1 + 1) + B(2(-1) - 1)(-1 + 1) + C(2(-1) - 1)(-1 + 4)$

$6 + 24 - 3 = A(3)(0) + B(-2 - 1)(0) + C(-2 - 1)(3)$

$27 = 0 + 0 + C(-3)(3)$

$27 = -9C$

$C = \dfrac{27}{-9}$

$C = -3$

Since we already know B and C we can substitute any value for x in the identity and obtain A.

Let us substitute 1 for x in order to obtain A. This gives:

$6x^2 - 24x - 3 \equiv A(x + 4)(x + 1) + B(2x - 1)(x + 1) + C(2x - 1)(x + 4)$

$6(1)^2 - 24(1) - 3 = A(1 + 4)(1 + 1) + B(2(1) - 1)(1 + 1) + C(2(1) - 1)(1 + 4)$

$6 - 24 - 3 = A(5)(2) + B(2 - 1)(2) + C(2 - 1)(5)$

$-21 = 10A + 2B + 5C$

$-21 = 10A + 2(7) + 5(-3)$ (Note that B = 7 and C = −3)

$-21 = 10A + 14 - 15$

$-21 = 10A - 1$

$-21 + 1 = 10A$

$-20 = 10A$

$A = \dfrac{-20}{10}$

$A = -2$

Hence A = −2, B = 7 and C = −3

Therefore, $\dfrac{6x^2 - 24x - 3}{(2x - 1)(x^2 + 5x + 4)} = -\dfrac{2}{2x - 1} + \dfrac{7}{x + 4} - \dfrac{3}{x + 1}$

Or, $\dfrac{6x^2 - 24x - 3}{(2x - 1)(x^2 + 5x + 4)} = \dfrac{7}{x + 4} - \dfrac{2}{2x - 1} - \dfrac{3}{x + 1}$

11. Express $\dfrac{6x^2 + 19x - 11}{(x + 1)(x^2 + 5x - 2)}$ in partial fractions

Solution

The quadratic factor in the denominator cannot be factorized into linear factors. Therefore we resolve into partial fractions as follows:

$$\frac{6x^2 + 19x - 11}{(x+1)(x^2+5x-2)} = \frac{A}{x+1} + \frac{Bx+C}{(x^2+5x-2)}$$

$$\frac{6x^2 + 19x - 11}{(x+1)(x^2+5x-2)} = \frac{A(x^2+5x-2)+(Bx+C)(x+1)}{(x+1)(x^2+5x-2)}$$

Equating the numerators gives:

$$6x^2 + 19x - 11 \equiv A(x^2 + 5x - 2) + (Bx + C)(x + 1)$$

Substitute -1 for x in order to obtain A as follows:

$$6(-1)^2 + 19(-1) - 11 = A((-1)^2 + 5(-1) - 2) + (B(-1) + C)(-1 + 1)$$

$$6 - 19 - 11 = A(1 - 5 - 2) + (-B + C)(0)$$

$$-24 = -6A + 0$$

$$6A = 24$$

$$A = \frac{24}{6}$$

$$A = 4$$

Since we have a quadratic factor in this question, we will have to apply the method of comparing coefficients. Hence, let us expand the brackets in the identity above. This gives:

$$6x^2 + 19x - 11 \equiv A(x^2 + 5x - 2) + (Bx + C)(x + 1)$$

$$6x^2 + 19x - 11 \equiv Ax^2 + 5Ax - 2A + Bx^2 + Bx + Cx + C$$

$$\equiv Ax^2 + Bx^2 + 5Ax + Bx + Cx - 2A + C$$

$$6x^2 + 19x - 11 \equiv (A + B)x^2 + (5A + B + C)x - 2A + C$$

Comparing the coefficients of like terms on both sides of the equation shows that:

A + B = 6 .... Equation (1)

5A + B + C = 19 ....Equation (2)

We have already calculated the value of A to be 4 as done above. Hence from equation (1) we have:

A + B = 6 ....Equation (1)

4 + B = 6   (Since A = 4)

B = 6 − 4

B = 2

From equation (2) we have:

5A + B + C = 19 ....Equation (2)

5(4) + 2 + C = 19

20 + 2 + C = 19

C = 19 − 22

C = −3

Hence, A = 4, B = 2 and C = −3

$$\frac{6x^2 + 19x - 11}{(x+1)(x^2+5x-2)} = \frac{A}{x+1} + \frac{Bx+C}{(x^2+5x-2)}$$

Therefore, $\dfrac{6x^2 + 19x - 11}{(x+1)(x^2+5x-2)} = \dfrac{4}{x+1} + \dfrac{2x-3}{(x^2+5x-2)}$

12. Resolve $\dfrac{2x^2 - 5x + 1}{(x-2)^3}$ into partial fractions

Solution

This is a case of repeated linear factor. Hence we resolve into partial fractions as follows:

$$\frac{2x^2 - 5x + 1}{(x-2)^3} = \frac{A}{(x-2)} + \frac{B}{(x-2)^2} + \frac{C}{(x-2)^3}$$

$$\frac{2x^2 - 5x + 1}{(x-2)^3} = \frac{A(x-2)^2 + B(x-2) + C}{(x-2)^3}$$

Equating the numerators gives:

$$2x^2 - 5x + 1 \equiv A(x-2)^2 + B(x-2) + C$$

Substitute 2 for x in order to eliminate A and B and obtain the value of C. This gives:

$2(2)^2 - 5(2) + 1 = A(2-2)^2 + B(2-2) + C$

$2(4) - 10 + 1 = A(0) + B(0) + C$

$8 - 10 + 1 = 0 + 0 + C$

$-1 = C$

$\quad C = -1$

In order to find A and B, we need to compare coefficients. Let us expand the brackets in the identity above as follows:

$$2x^2 - 5x + 1 \equiv A(x-2)^2 + B(x-2) + C$$
$$\equiv A(x-2)(x-2) + B(x-2) + C$$
$$\equiv A(x^2 - 4x + 4) + Bx - 2B + C$$
$$\equiv Ax^2 - 4Ax + 4A + Bx - 2B + C$$
$$2x^2 - 5x + 1 \equiv Ax^2 + (-4A + B)x + 4A - 2B + C$$

Comparing the coefficients of like terms on both sides of the equation shows that:

$\quad A = 2 \quad$ (Since $2x^2 \equiv Ax^2$)

Similarly, $\quad -4A + B = -5 \quad$ (From $-5x \equiv (-4A + B)x$)

$\qquad -4(2) + B = -5 \quad$ (Note that A = 2 as obtained above)

$\qquad -8 + B = -5$

$\qquad B = -5 + 8$

$\qquad B = 3$

Hence, A = 2, B = 3 and C = −1

Therefore, $\dfrac{2x^2 - 5x + 1}{(x-2)^3} = \dfrac{2}{(x-2)} + \dfrac{3}{(x-2)^2} - \dfrac{1}{(x-2)^3}$

13. Resolve $\dfrac{5x^3+3x^2+2x-1}{x^2(x^2+1)}$ into partial fractions

Solution

Let $\dfrac{5x^3+3x^2+2x-1}{x^2(x^2+1)} = \dfrac{A}{x} + \dfrac{B}{x^2} + \dfrac{Cx+D}{x^2+1}$

$\dfrac{5x^3+3x^2+2x-1}{x^2(x^2+1)} = \dfrac{A[x(x^2+1)]+B(x^2+1)+(Cx+D)x^2}{x^2(x^2+1)}$

Equating the numerators gives:

$5x^3+3x^2+2x-1 \equiv A[x(x^2+1)]+B(x^2+1)+(Cx+D)(x^2)$

$\equiv A(x^3+x)+B(x^2+1)+Cx^3+Dx^2$

$\equiv Ax^3+Ax+Bx^2+B+Cx^3+Dx^2$

$\equiv Ax^3+Cx^3+Bx^2+Dx^2+Ax+B$

$5x^3+3x^2+2x-1 \equiv (A+C)x^3+(B+D)x^2+Ax+B$

Comparing the coefficients of terms on both sides of the identity, gives:

$\quad A = 2$ (From coefficients of x)

$\quad B = -1$ (The constant term on both sides of the identity)

$\quad A + C = 5$ (From coefficients of x^3)

$\quad 2 + C = 5$ (Since A = 2)

$\quad C = 5 - 2$

$\quad C = 3$

Similarly, $B + D = 3$ (From coefficients of x^2)

$\quad\quad -1 + D = 3$

$\quad\quad D = 3 + 1$

$\quad\quad D = 4$

Hence, A = 2, B = −1, C = 3 and D = 4

Therefore, $\dfrac{5x^3+3x^2+2x-1}{x^2(x^2+1)} = \dfrac{2}{x} - \dfrac{1}{x^2} + \dfrac{3x+4}{x^2+1}$

14. Resolve $\dfrac{2x^2+19x+47}{x^2+9x+20}$ into partial fractions.

Solutions

From the question the highest power of x in the numerator and denominator are equal (i.e. x^2 in both cases). Hence we have to divide the numerator by the denominator in order to reduce the power of x in the numerator. Let us divide it as follows:

$$
\begin{array}{r}
2 \\
x^2 + 9x + 20\overline{\smash{\big)}\,2x^2 + 19x + 47} \\
\underline{2x^2 + 18x + 40} \\
x + 7
\end{array}
$$

We cannot divide further, hence:

$$\frac{2x^2 + 19x + 47}{x^2 + 9x + 20} = 2 + \frac{x + 7}{x^2 + 9x + 20}$$

Note that 2 is the quotient and $x + 7$ is the remainder in the division carried out above.

Hence we now resolve $\dfrac{x + 7}{x^2 + 9x + 20}$ into partial fractions.

The denominator factorizes into linear factors as follows:

$$x^2 + 9x + 20 = (x + 4)(x + 5)$$

Hence the fraction above can be written as:

$$\frac{x + 7}{(x+4)(x+5)}$$

Therefore, $\dfrac{x + 7}{(x+4)(x+5)} = \dfrac{A}{(x+4)} + \dfrac{B}{(x+5)}$

$$\frac{x + 7}{(x+4)(x+5)} = \frac{A(x+5) + B(x+4)}{(x+4)(x+5)}$$

Equating the numerators gives:

$$x + 7 \equiv A(x + 5) + B(x + 4)$$

Substitute $x = -5$ in order to get B. This gives:

$$-5 + 7 = A(-5 + 5) + B(-5 + 4)$$
$$2 = A(0) + B(-1)$$
$$2 = 0 - B$$
$$B = -2$$

Substitute $x = -4$ in order to find A. This gives:

$$x + 7 \equiv A(x + 5) + B(x + 4)$$
$$-4 + 7 = A(-4 + 5) + B(-4 + 4)$$
$$3 = A(1) + B(0)$$
$$3 = A + 0$$
$$A = 3$$

Hence A = 3 and B = −2

Therefore, $\dfrac{2x^2 + 19x + 47}{x^2 + 9x + 20} = 2 + \dfrac{x + 7}{x^2 + 9x + 20}$ (Note: 2 was obtained from the division above)

$$= 2 + \frac{3}{(x+4)} - \frac{2}{(x+5)}$$

15. Resolve $\dfrac{x^3 + 2x^2 - x - 11}{x^2 - x - 2}$ into partial fractions

Solution

From the question, the highest power of x in the numerator is higher than the highest power of x in the denominator (i.e. x^3 is greater than x^2). In such a case, we have to divide first before we proceed. Let us carry out the division as follows:

146

$$\begin{array}{r} x + 3 \\ x^2 - x - 2 \overline{)\, x^3 + 2x^2 - x - 11} \\ \underline{x^3 - x^2 - 2x } \\ 3x^2 + x - 11 \\ \underline{3x^2 - 3x - 6} \\ 4x - 5 \end{array}$$

Hence, using the remainder which is $4x - 5$, the fraction in our question can be written as:

$$\frac{x^3 + 2x^2 - x - 11}{x^2 - x - 2} = x + 3 + \frac{4x - 5}{x^2 - x - 2}$$

We can now resolve $\dfrac{4x - 5}{x^2 - x - 2}$ into partial fractions. The denominator can be factorized to give:

$x^2 - x - 2 = (x - 2)(x + 1)$. Hence, the fraction to resolve can be written as $\dfrac{4x - 5}{(x - 2)(x + 1)}$

Therefore we resolve as follows:

$$\frac{4x - 5}{(x - 2)(x + 1)} = \frac{A}{x - 2} + \frac{B}{x + 1}$$

$$\frac{4x - 5}{(x - 2)(x + 1)} = \frac{A(x + 1) + B(x - 2)}{(x - 2)(x + 1)}$$

Equating the numerators gives:

$$4x - 5 \equiv A(x + 1) + B(x - 2)$$

Substitute $x = -1$ in order to get B. This gives:

$$4(-1) - 5 = A(-1 + 1) + B(-1 - 2)$$

$$-4 - 5 = A(0) + B(-3)$$

$$-9 = 0 - 3B$$

$$3B = 9$$

$$B = \frac{9}{3}$$

$$B = 3$$

Substitute $x = 2$ in order to find A. This gives:

$$4x - 5 \equiv A(x + 1) + B(x - 2)$$

$$4(2) - 5 = A(2 + 1) + B(2 - 2)$$

$$8 - 5 = A(3) + B(0)$$

$$3 = 3A + 0$$

$$A = \frac{3}{3}$$

$$A = 1$$

Hence A = 1 and B = 3

Therefore, $\dfrac{x^3 + 2x^2 - x - 11}{x^2 - x - 2} = x + 3 + \dfrac{1}{x - 2} + \dfrac{3}{x + 1}$

16. If $\dfrac{x^2+x-1}{(x+1)(x-1)} = A + \dfrac{B}{x+1} + \dfrac{C}{x-1}$ where A, B and C are constants, find $A + 2B - C$

Solution

If we expand the denominators, it gives:

$$(x+1)(x-1) = x^2 - x + x - 1$$
$$= x^2 - 1$$

This shows that the numerator and the denominator are of the same degree (i.e. both have the same highest power of x). Hence we have to divide first. This is as shown below.

$$\begin{array}{r} 1 \\ x^2 - 1 \overline{)x^2 + x - 1} \\ \underline{x^2 + 0x - 1} \\ x \end{array}$$

(Note that $0x$ is added since there is no term in x)

Therefore with x as the remainder, we now write the fraction as follows:

$$\frac{x^2+x-1}{(x+1)(x-1)} = 1 + \frac{x}{(x+1)(x-1)}$$ (This shows that A = 1)

Let us now resolve $\dfrac{x}{(x+1)(x-1)}$ into partial fractions as follows.

$$\frac{x}{(x+1)(x-1)} = \frac{B}{x+1} + \frac{C}{x-1}$$

$$\frac{x}{(x+1)(x-1)} = \frac{B(x-1) + C(x+1)}{(x+1)(x-1)}$$

Equating the numerators gives:

$$x \equiv B(x-1) + C(x+1)$$

If we put $x = 1$, we obtain C as follows:

$$1 = B(1-1) + C(1+1)$$
$$1 = B(0) + C(2)$$
$$1 = 0 + 2C$$
$$1 = 2C$$
$$C = \frac{1}{2}$$

If we put $x = -1$, we obtain B as follows:

$$-1 = B(-1-1) + C(-1+1)$$
$$-1 = B(-2) + C(0)$$
$$-1 = -2B + 0$$
$$2B = 1$$
$$B = \frac{1}{2}$$

Hence, $B = \dfrac{1}{2}$ and $C = \dfrac{1}{2}$. And recall that A = 1

Therefore, $A + 2B - C = 1 + 2\left(\frac{1}{2}\right) - \frac{1}{2}$

$$= 1 + 1 - \frac{1}{2}$$

$$= 2 - \frac{1}{2}$$

$$= 1\frac{1}{2}$$

17. If $\dfrac{3x^2 - 7}{x^3 + 2x^2 - 8x} \equiv \dfrac{7}{8x} + \dfrac{P}{x + 4} + \dfrac{Q}{x - 2}$, find $P + Q$

Solution

Looking at the denominators on both sides of the identity shows that the factors of $x^3 + 2x^2 - 8x$ are x, $(x + 4)$ and $(x - 2)$. This means that:

$$x^3 + 2x^2 - 8x = x\,(x + 4)(x - 2)$$

This shows that $\dfrac{7}{8x}$ should be written as $\dfrac{\frac{7}{8}}{x}$. If we use $\dfrac{7}{8x}$, then the denominators on both sides of the identity will not be identical. For the two denominators to be identical, we must write the question above as follows:

$$\frac{3x^2 - 7}{x^3 + 2x^2 - 8x} \equiv \frac{\frac{7}{8}}{x} + \frac{P}{x + 4} + \frac{Q}{x - 2}$$

We now continue our working as follows:

$$\frac{3x^2 - 7}{x(x + 4)(x - 2)} \equiv \frac{\frac{7}{8}(x + 4)(x - 2) + P(x)(x - 2) + Q(x)(x + 4)}{x(x + 4)(x - 2)}$$

Equating the numerators gives:

$$3x^2 - 7 \equiv \frac{7}{8}(x + 4)(x - 2) + P(x)(x - 2) + Q(x)(x + 4)$$

Substitute -4 for x to obtain P as follows

$$3(-4)^2 - 7 \equiv \frac{7}{8}(-4 + 4)(-4 - 2) + P(-4)(-4 - 2) + Q(-4)(-4 + 4)$$

$$3(16) - 7 = \frac{7}{8}(0)(-6) + P(-4)(-6) + Q(-4)(0)$$

$$48 - 7 = 0 + 24P + 0$$

$$41 = 24P$$

$$P = \frac{41}{24}$$

Let us put $x = 2$ in order to get Q.

$$3x^2 - 7 \equiv \frac{7}{8}(x + 4)(x - 2) + P(x)(x - 2) + Q(x)(x + 4)$$

$$3(2)^2 - 7 \equiv \frac{7}{8}(2 + 4)(2 - 2) + P(2)(2 - 2) + Q(2)(2 + 4)$$

$$3(4) - 7 = \frac{7}{8}(6)(0) + P(2)(0) + Q(2)(6)$$

$$12 - 7 = 0 + 0 + 12Q$$

$$5 = 12Q$$

$$Q = \frac{5}{12}$$

Therefore, $P = \frac{41}{24}$, and $Q = \frac{5}{12}$

Hence, $P + Q = \frac{41}{24} + \frac{5}{12}$

$$= \frac{41 + 10}{24}$$

$$= \frac{51}{24}$$

$$\therefore \quad P + Q = \frac{17}{8} \qquad \text{(In its lowest term)}$$

18. Resolve $\dfrac{1 - 4x + 7x^2 - x^3 - x^4}{(x + 3)(x^2 + 2)}$ into partial fractions.

Solutions

If we expand the denominator, it will give us a degree (highest power of x) of 3 (i.e. x^3). However the numerator contains x^4. Hence the fraction is an improper fraction. Therefore we have to divide the fraction before we proceed.

Let us expand the denominator as follows:

$$(x + 3)(x^2 + 2) = x^3 + 2x + 3x^2 + 6$$

$$= x^3 + 3x^2 + 2x + 6$$

We also have to rearrange the numerator so that the powers of x will be in descending order. This is necessary for easier division of the polynomial. Hence the fraction can be written as follows:

$$\frac{-x^4 - x^3 + 7x^2 - 4x + 1}{x^3 + 3x^2 + 2x + 6}$$

Let us now carry out the division as follows:

$$
\begin{array}{r}
-x + 2 \\
x^3 + 3x^2 + 2x + 6 \overline{) -x^4 - x^3 + 7x^2 - 4x + 1} \\
\underline{-x^4 - 3x^3 - 2x^2 - 6x} \\
2x^3 + 9x^2 + 2x + 1 \\
\underline{2x^3 + 6x^2 + 4x + 12} \\
3x^2 - 2x - 11
\end{array}
$$

Note that we can no longer continue the division when there is no more term to bring down and the highest power of x in the remainder is smaller than that of the divisor. Hence, the fraction has been broken down to give:

$$\frac{1 - 4x + 7x^2 - x^3 - x^4}{(x + 3)(x^2 + 2)} = 2 - x + \frac{3x^2 - 2x - 11}{(x + 3)(x^2 + 2)} \qquad \text{(Note that } -x + 2 \text{ has been written as } 2 - x\text{)}$$

150

Let us now resolve $\dfrac{3x^2 - 2x - 11}{(x+3)(x^2+2)}$ into partial fractions as follows:

$$\frac{3x^2 - 2x - 11}{(x+3)(x^2+2)} = \frac{A}{x+3} + \frac{Bx+C}{(x^2+2)}$$

$$\frac{3x^2 - 2x - 11}{(x+3)(x^2+2)} = \frac{A(x^2+2) + (Bx+C)(x+3)}{(x+3)(x^2+2)}$$

Equating the numerators gives:

$$3x^2 - 2x - 11 \equiv A(x^2+2) + (Bx+C)(x+3)$$

Substitute −3 for x in order to find A. This gives:

$$3(-3)^2 - 2(-3) - 11 = A((-3)^2+2) + (B(-3)+C)(-3+3)$$

$$3(9) + 6 - 11 = A(9+2) + (-3B+C)(0)$$

$$27 + 6 - 11 = 11A + 0$$

$$22 = 11A$$

$$A = \frac{22}{11}$$

$$A = 2$$

Let us expand the brackets in the identity above.

$$3x^2 - 2x - 11 \equiv A(x^2+2) + (Bx+C)(x+3)$$

$$\equiv Ax^2 + 2A + Bx^2 + 3Bx + Cx + 3C$$

$$\equiv Ax^2 + Bx^2 + 3Bx + Cx + 2A + 3C$$

$$3x^2 - 2x - 11 \equiv (A+B)x^2 + (3B+C)x + 2A + 3C$$

Comparing coefficients of like terms on both sides of the identity shows that:

$\qquad$ A + B = 3 ....Equation (1)

$\qquad$ 3B + C = −2 ....Equation (2)

Since we have obtained the value of A as 2, let us substitute this in equation (1) as follows:

$\qquad$ A + B = 3 ....Equation (1)

$\qquad$ 2 + B = 3 $\qquad$ (Note that A = 2 as obtained above)

$\qquad$ B = 3 − 2

$\qquad$ B = 1

From equation (2) we have that:

$\qquad$ 3B + C = −2 ....Equation (2)

$\qquad$ 3(1) + C = −2

$\qquad$ 3 + C = −2

$\qquad$ C = −2 − 3

$\qquad$ C = −5

Hence, A = 2, B = 1 and C = −5

Therefore, $\dfrac{1 - 4x + 7x^2 - x^3 - x^4}{(x+3)(x^2+2)} = 2 - x + \dfrac{2}{x+3} + \dfrac{x-5}{(x^2+2)}$

Exercise 10

1. Express $\dfrac{5x - 1}{(x + 3)(x - 5)}$ in partial fractions

2. Resolve $\dfrac{2x + 11}{2x^2 - 3x - 9}$ into partial fractions.

3. Express $\dfrac{8}{x^2 - 16}$ in partial fractions.

4. Resolve $\dfrac{3x - 10}{2x(x - 5)}$ into partial fractions.

5. Resolve $\dfrac{x + 7}{(x - 2)^2}$ into partial fractions.

6. Express $\dfrac{6 + 4x - 5x^2}{x^2(2x + 3)}$ in partial fractions.

7. Resolve $\dfrac{2x^2 - 5x + 11}{(x + 1)(x - 2)^2}$ into partial fractions.

8. Resolve $\dfrac{5x^2 - x + 2}{x(x^2 + 3)}$ into partial fractions.

9. Resolve $\dfrac{3x^2 + 2x + 8}{(x^2 + 2)(x + 1)}$ into partial fractions

10. Resolve $\dfrac{23x - 31}{(x - 2)(x^2 + 2x - 3)}$ into partial fractions

11. Express $\dfrac{5x^2 + 2x - 9}{(x + 3)(x^2 - 3x - 3)}$ in partial fractions

12. Resolve $\dfrac{x^2 + x - 7}{(x - 1)^3}$ into partial fractions

13. Resolve $\dfrac{2x^3 + 3x^2 - x - 5}{x^2(x^2 - 1)}$ into partial fractions

14. Resolve $\dfrac{x^2 - 3x - 1}{x^2 - 2x - 3}$ into partial fractions.

15. Resolve $\dfrac{x^3 + 8x^2 + 13x - 29}{x^2 + 3x - 4}$ into partial fractions

16. If $\dfrac{5x^2 + 13x - 14}{(x - 1)(x + 3)} = A + \dfrac{B}{x - 1} + \dfrac{C}{x + 3}$ where A, B and C are constants, find 2A + 5B − 7C

17. If $\dfrac{2x^2 + 5}{2x^3 + 14x^2 + 20x} \equiv \dfrac{1}{2x} + \dfrac{P}{x + 5} + \dfrac{Q}{x + 2}$, find the values of P and Q

18. Resolve $\dfrac{4 - 2x + 5x^2 + 2x^3 - x^4}{(x - 1)(x^2 + 1)}$ into partial fractions.

152

CHAPTER 11
RADICAL EQUATIONS

Equations which consist of square root signs are called radical equations. In solving such equations, we keep squaring both sides of the equation until the square root sign is finally removed. Then we solve the resulting equation after the removal of the root sign.

However, it is important that we recall the following rules when simplifying surds:

1. $\sqrt{a} \times \sqrt{b} = \sqrt{ab}$

2. $\dfrac{\sqrt{a}}{\sqrt{b}} = \sqrt{\dfrac{a}{b}}$

3. $\sqrt{a} \times \sqrt{a} = a$

Examples

1. Solve the equation $\sqrt{x - 2} = 3$

Solution

$\sqrt{x - 2} = 3$

Square both sides in order to remove the root sign. This gives:

$(\sqrt{x - 2})^2 = 3^2$

$x - 2 = 9$

Note that the expression $(\sqrt{x - 2})^2$ gives $x - 2$, i.e. simply take the term in the root sign.

$x = 9 + 2$

$x = 11$

2. Solve $\sqrt{x + 14} = \sqrt{6 - x}$

Solution

$\sqrt{x + 14} = \sqrt{6 - x}$

Squaring both sides gives:

$(\sqrt{x + 14})^2 = (\sqrt{6 - x})^2$

$x + 14 = 6 - x$

$x + x = 6 - 14$

$2x = -8$

$x = \dfrac{-8}{2}$

$x = -4$

3. Solve the equation $\sqrt{x} + \sqrt{2x - 2} = 7$

Solution

$$\sqrt{x} + \sqrt{2x - 2} = 7$$

Square both sides of the equation in order to remove the root sign. This gives:

$$(\sqrt{x} + \sqrt{2x - 2})^2 = 7^2$$
$$(\sqrt{x} + \sqrt{2x - 2})(\sqrt{x} + \sqrt{2x - 2}) = 49$$

In order to expand the left hand side, simply use each term in the first bracket to multiply the second bracket. This is as shown below.

$$(\sqrt{x})(\sqrt{x} + \sqrt{2x - 2}) + \sqrt{2x - 2}(\sqrt{x} + \sqrt{2x - 2}) = 49$$
$$x + \sqrt{x}\sqrt{2x - 2} + \sqrt{x}\sqrt{2x - 2} + 2x - 2 = 49$$
$$x + 2x - 2 + \sqrt{x(2x - 2)} + \sqrt{x(2x - 2)} = 49$$
$$3x - 2 + 2\sqrt{x(2x - 2)} = 49 \qquad [\text{Note that } \sqrt{x(2x - 2)} + \sqrt{x(2x - 2)} = 2\sqrt{x(2x - 2)}]$$
$$3x + 2\sqrt{2x^2 - 2x} = 49 + 2$$
$$3x + 2\sqrt{2x^2 - 2x} = 51$$
$$2\sqrt{2x^2 - 2x} = 51 - 3x$$

Now that we have a single root sign, we have to square both sides of the equation to remove the root sign again. This gives:

$$(2\sqrt{2x^2 - 2x})^2 = (51 - 3x)^2$$
$$(2)^2(\sqrt{2x^2 - 2x})^2 = (51 - 3x)(51 - 3x)$$
$$4(2x^2 - 2x) = 2601 - 153x - 153x + 9x^2$$
$$8x^2 - 8x = 2601 - 306x + 9x^2$$
$$0 = 9x^2 - 8x^2 + 8x - 306x + 2601$$
$$0 = x^2 - 298x + 2601$$

Or, $x^2 - 298x + 2601$

Solving this equation by factorization gives:

$$(x - 9)(x - 289) = 0$$
$$\therefore \quad x = 9 \text{ or } x = 289$$

After solving a radical equation it is necessary to check if the values obtained are actually the true solutions of the equation. Hence, let us substitute each of the values of x obtained above into the equation. The equation is:

$$\sqrt{x} + \sqrt{2x - 2} = 7$$

When $x = 9$, the left hand side of the equation gives:

$$\sqrt{9} + \sqrt{2(9) - 2} = 3 + \sqrt{18 - 2}$$
$$= 3 + \sqrt{16}$$
$$= 3 + 4$$
$$= 7$$

Since 7 is also the value on the right hand side of the equation, then $x = 9$ is a solution of the equation.

Recall that the equation is $\sqrt{x} + \sqrt{2x - 2} = 7$

When $x = 289$, the left hand side of the equation gives:

$$\sqrt{289} + \sqrt{2(289) - 2} = 17 + \sqrt{578 - 2}$$
$$= 17 + \sqrt{576}$$
$$= 17 + 24$$
$$= 41$$

Since 41 is not what we have on the right hand side of the equation, then $x = 289$ is not a solution of the equation.

Therefore, the solution of the equation is $x = 9$

Note that 289 which is not a solution of the equation is called an extraneous root of the equation.

4. Solve the equation $\sqrt{3x + 13} - x = 1$

Solution

$$\sqrt{3x + 13} - x = 1$$

Collect terms containing the root sign (radical) on one side of the equation, and move other terms to the other side of the equation. This means we have to isolate the radical term. This gives:

$$\sqrt{3x + 13} = 1 + x$$

We can now square both sides of the equation as follows:

$$(\sqrt{3x + 13})^2 = (1 + x)^2$$
$$3x + 13 = (1 + x)(1 + x)$$
$$3x + 13 = 1 + x + x + x^2$$
$$3x + 13 = x^2 + 2x + 1$$
$$0 = x^2 + 2x - 3x + 1 - 13$$
$$0 = x^2 - x - 12$$
$$x^2 - x - 12 = 0$$

Solving this equation by factorization gives:

$$(x - 4)(x + 3) = 0$$

Hence, $x = 4$ or $x = -3$

Let us check the values of x obtained in order to know if they are solutions of the equation. The equation is:

$$\sqrt{3x + 13} - x = 1$$

When $x = 4$, the left hand side of the equation gives:

$$\sqrt{3x + 13} - x = \sqrt{3(4) + 13} - 4$$

155

$$= \sqrt{12 + 13} - 4$$

$$= \sqrt{25} - 4$$

$$= 5 - 4 = 1$$

Since 1 is also the value on the right hand side of the equation, the $x = 4$ is a solution of the equation.

When $x = -3$, the left hand side of the equation gives:

$$\sqrt{3x + 13} - x = \sqrt{3(-3) + 13} - (-3)$$

$$= \sqrt{-9 + 13} + 3$$

$$= \sqrt{4} + 3$$

$$= 2 + 3$$

$$= 5$$

Since 5 is not what we have on the right hand side of the equation, then $x = -3$ is not a solution of the equation.

Therefore, the solution of the equation is $x = 4$

5. Solve for x if $\sqrt{x} + 2\sqrt{x + 8} = 7$

Solution

$$\sqrt{x} + 2\sqrt{x + 8} = 7$$

Squaring both sides of the equation gives:

$$(\sqrt{x} + 2\sqrt{x + 8})^2 = 7^2$$

$$(\sqrt{x} + 2\sqrt{x + 8})(\sqrt{x} + 2\sqrt{x + 8}) = 49$$

$$x + 2\sqrt{x(x + 8)} + 2\sqrt{x(x + 8)} + 4(x + 8) = 49$$

$$x + 4\sqrt{x(x + 8)} + 4x + 32 = 49$$

$$4\sqrt{x^2 + 8x} = 49 - 32 - 4x - x$$

$$4\sqrt{x^2 + 8x} = 17 - 5x$$

Squaring both sides again gives:

$$(4\sqrt{x^2 + 8x})^2 = (17 - 5x)^2$$

$$4^2(\sqrt{x^2 + 8x})^2 = (17 - 5x)(17 - 5x)$$

$$16(x^2 + 8x) = 289 - 85x - 85x + 25x^2$$

$$16x^2 + 128x = 289 - 170x + 25x^2$$

$$0 = 25x^2 - 16x^2 - 170x - 128x + 289$$

$$0 = 9x^2 - 298x + 289$$

$$9x^2 - 298x + 289 = 0$$

Let us solve this equation by factorization (Note that quadratic formula can also be used) as follows:

$$9x^2 - 9x - 289x + 289 = 0$$

156

$9x(x-1) - 289(x-1) = 0$

$(x-1)(9x-289) = 0$

Hence, $x = 1$ or $x = \dfrac{289}{9}$

$\dfrac{289}{9}$ cannot be the solution to the equation since it is a fraction and it is too large (the right hand side is just 7).

Therefore the solution to the equation is $x = 1$

6. Solve for x if $\sqrt{x+7} + 3x = 9$

Solution

$\sqrt{x+7} + 3x = 9$

Collect term containing the radical on the left hand side of the equation. This gives:

$\sqrt{x+7} = 9 - 3x$

Square both sides of the equation

$(\sqrt{x+7})^2 = (9-3x)^2$

$x + 7 = (9-3x)(9-3x)$

$x + 7 = 81 - 27x - 27x + 9x^2$

$0 = 81 - 54x + 9x^2 - x - 7$

$0 = 9x^2 - 55x + 74$

$9x^2 - 55x + 74 = 0$

Let us solve this equation by using quadratic formula. From the equation:

$a = 9$, $b = -55$ and $c = 74$

$x = \dfrac{-b \pm \sqrt{b^2 - 4ac}}{2a}$

$= \dfrac{-(-55) \pm \sqrt{(-55)^2 - (4 \times 9 \times 74)}}{2 \times 9}$

$= \dfrac{55 \pm \sqrt{3025 - 2664}}{18}$

$= \dfrac{55 \pm \sqrt{361}}{18}$

$= \dfrac{55 \pm 19}{18}$

$x = \dfrac{55 + 19}{18}$ or $x = \dfrac{55 - 19}{18}$

$x = \dfrac{74}{18}$ or $x = \dfrac{36}{18}$

$x = \dfrac{37}{9}$ or $x = 2$

$\dfrac{37}{9}$ cannot be the solution of the equation if substituted into the equation.

Therefore the solution to the equation is $x = 2$

7. Find the value of x in the equation $\sqrt{x+1} + \sqrt{x+6} = 5$

Solution

$$\sqrt{x+1} + \sqrt{x+6} = 5$$

Squaring both sides of the equation gives:

$$(\sqrt{x+1} + \sqrt{x+6})^2 = 5^2$$

$$(\sqrt{x+1} + \sqrt{x+6})(\sqrt{x+1} + \sqrt{x+6}) = 25$$

$$(\sqrt{x+1} + \sqrt{x+6})(\sqrt{x+1} + \sqrt{x+6}) = 25$$

$$x + 1 + \sqrt{(x+1)(x+6)} + \sqrt{(x+1)(x+6)} + x + 6 = 25$$

$$x + 1 + 2\sqrt{(x+1)(x+6)} + x + 6 = 25$$

$$2x + 7 + 2\sqrt{(x+1)(x+6)} = 25$$

Isolating the radical term on the left hand side gives:

$$2\sqrt{(x+1)(x+6)} = 25 - 2x - 7$$

$$2\sqrt{x^2 + 7x + 6} = 18 - 2x$$

Divide both sides by 2. This gives:

$$\sqrt{x^2 + 7x + 6} = 9 - x$$

Squaring both sides of the equation gives:

$$(\sqrt{x^2 + 7x + 6})^2 = (9-x)^2$$

$$x^2 + 7x + 6 = (9-x)(9-x)$$

$$x^2 + 7x + 6 = 81 - 18x + x^2$$

$$x^2 - x^2 + 7x + 18x = 81 - 6$$

$$25x = 75$$

$$x = \frac{75}{25}$$

$$x = 3$$

8. Find the value of x if $\sqrt{2x^2 + 7} = 3$

Solution

$$\sqrt{2x^2 + 7} = 3$$

Squaring both sides of the equation gives:

$$(\sqrt{2x^2 + 7})^2 = 3^2$$

$$2x^2 + 7 = 9$$

$$2x^2 = 9 - 7$$

$$2x^2 = 2$$

$$x^2 = 1$$

$$x = \sqrt{1}$$

$$x = \pm 1$$

$x = 1$ or $x = -1$

Let us check if the two values of x are solutions of the equation. The equation is:

$\sqrt{2x^2 + 7} = 3$

When $x = 1$, the left hand side of the equation gives:

$$\sqrt{2x^2 + 7} = \sqrt{2(1)^2 + 7}$$
$$= \sqrt{2 + 7}$$
$$= \sqrt{9}$$
$$= 3$$

Hence $x = 1$ is a solution of the equation.

When $x = -1$, the left hand side of the equation gives:

$$\sqrt{2x^2 + 7} = \sqrt{2(-1)^2 + 7}$$
$$= \sqrt{2(1) + 7}$$
$$= \sqrt{2 + 7}$$
$$= \sqrt{9} = 3$$

Hence $x = -1$ is a solution of the equation.

Therefore both $x = 1$, and $x = -1$ are solutions of the equation.

9. Solve the equation $\sqrt{5x^2 - 2x} = 4$

Solution

$\sqrt{5x^2 - 2x} = 4$

Squaring both sides of the equation gives:

$$(\sqrt{5x^2 - 2x})^2 = 4^2$$
$$5x^2 - 2x = 16$$
$$5x^2 - 2x - 16 = 0$$

Solving this equation by factorization gives:

$$5x^2 - 10x + 8x - 16 = 0$$
$$5x(x - 2) + 8(x - 2) = 0$$
$$(x - 2)(5x + 8) = 0$$
$$x = 2 \quad \text{or} \quad x = -\frac{8}{5}$$

If we substitute $x = 2$ and $x = -\frac{8}{5}$ into the equation, we will find out that the two values of x are solutions of the equation.

Therefore, both $x = 2$ and $x = -\frac{8}{5}$ are solutions of the equation.

10. Solve $\sqrt{x^2 - 9} - x = -1$

Solution

Take $-x$ to the other side of the equation in order to isolate the radical term. This gives:

$\sqrt{x^2 - 9} = x - 1$

Squaring both sides of the equation gives:

$$(\sqrt{x^2 - 9})^2 = (x - 1)^2$$
$$x^2 - 9 = x^2 - 2x + 1$$
$$x^2 - x^2 + 2x = 1 + 9$$
$$2x = 10$$
$$x = \frac{10}{2}$$
$$x = 5$$

Exercise 11

1. Solve the equation $\sqrt{2x - 3} = 7$
2. Solve $\sqrt{2x - 9} = \sqrt{x - 2}$
3. Solve the equation $\sqrt{2x} - \sqrt{4x - 2} = -1$
4. Solve the equation $\sqrt{2x + 16} - x = -4$
5. Solve for x if $\sqrt{4x} + 5\sqrt{x + 3} = 12$
6. Solve for x if $\sqrt{2x - 6} + 3x = 17$
7. Find the value of x in the equation $\sqrt{x - 1} + \sqrt{x + 4} = 5$
8. Find the value of x if $\sqrt{3x^2 + 13} = 11$
9. Solve the equation $\sqrt{2x^2 - 4x} = 4$
10. Solve $\sqrt{5x^2 + 4} - 2x = 1$

CHAPTER 12
LIMIT OF A FUNCTION

The concept of limits is very important in differential calculus. A function has a limiting value when its variable approaches a certain value. For example, if the limiting value of f(x) as x approaches 4 is 16, it is written as:

$$\lim_{x \to 4} f(x) = 16$$

In evaluating limit, some problems are simply solved by just putting in the values of the variables, while some problems are solved by applying certain rules.

Some important limits

1. $\lim\limits_{x \to 0} \dfrac{\sin x}{x} = 1$ and $\lim\limits_{x \to 0} \dfrac{\tan x}{x} = 1$ (Or $\lim\limits_{x \to 0} \dfrac{\sin ax}{ax} = 1$ and $\lim\limits_{x \to 0} \dfrac{\tan ax}{ax} = 1$)

2. $\lim\limits_{x \to 0} \dfrac{1 - \cos x}{x} = 0$ and $\lim\limits_{x \to 0} \dfrac{1 - \cos a x}{x} = 0$

3. $\lim\limits_{x \to \infty} \left(1 + \dfrac{1}{x}\right)^x = e$

4. $\lim\limits_{x \to 0} \dfrac{a^x - 1}{a^x} = 1$

5. $\lim\limits_{x \to 0} \dfrac{e^x - 1}{x} = 1$

6. $\lim\limits_{x \to 0} \dfrac{1 + x}{x} = 1$

7. $\lim\limits_{x \to a} \dfrac{x^n - a^n}{x - a} = na^{n-1}$

8. $\lim\limits_{x \to 0} \dfrac{a^x - 1}{x} = \ln a$

9. $\lim\limits_{x \to \infty} \dfrac{\ln x}{x} = 0$

10. $\lim\limits_{x \to \infty} x^{\frac{1}{x}} = 0$

11. $\lim\limits_{x \to 0} (1 + x)^{\frac{1}{x}} = e$

12. $\lim\limits_{x \to 0} (1 + \sin x)^{\frac{1}{x}} = e$

13. $\lim\limits_{x \to a} c = c$

Examples

1. Evaluate $\lim\limits_{x\to 0} 3x^3 - 5x^2 + 2x + 6$

<u>Solution</u>

$\lim\limits_{x\to 0} 3x^3 - 5x^2 + 2x + 6$

Simply substitute zero for x in the given expression. This gives:

$= 3(0^3) - 5(0^2) + 2(0) + 6$

$= 0 - 0 + 0 + 6$

$= 6$

2. Evaluate $\lim\limits_{x\to 0} \dfrac{x^2 + 3x - 11}{4x^2 - 5x - 5}$

<u>Solution</u>

$\lim\limits_{x\to 0} \dfrac{x^2 + 3x - 11}{4x^2 - 5x - 5}$

Substituting 0 for x into the expression gives:

$= \dfrac{0^2 + 3(0) - 11}{4(0)^2 - 5(0) - 5}$

$= \dfrac{0 + 0 - 11}{0 - 0 - 5}$

$= \dfrac{-11}{-5}$

$= \dfrac{11}{5}$

$= 2\dfrac{1}{5}$

3. Evaluate $\lim\limits_{x\to 4} \dfrac{x^2 - 2x - 8}{x - 4}$

<u>Solution</u>

A close look at the expression shows that if 4 is substituted for x, it will give $\dfrac{0}{0}$. This has no value as it is indeterminate. Therefore, in order to solve a limit such as this, we have to factorize the denominator and then simplify the expression. This is done as follows:

$\lim\limits_{x\to 4} \dfrac{x^2 - 2x - 8}{x - 4} = \lim\limits_{x\to 4} \dfrac{(x+2)(x-4)}{x - 4}$

Cancelling out $(x - 4)$ gives:

$\lim\limits_{x\to 4} (x + 2)$

We now substitute 4 for x to obtain our final answer as follows:

$\lim\limits_{x\to 4} (x + 2) = 4 + 2$

$= 6$

Therefore, $\lim\limits_{x\to 4} \dfrac{x^2 + x - 8}{x - 4} = 6$

4. Evaluate $\lim\limits_{x \to 2} \dfrac{x^2 - 4}{x - 2}$

Solution

This is similar to example 3 above since substituting 2 for x in the expression will $\dfrac{0}{0}$ which has no value. Hence we factorize the numerator and simplify as follows:

$\lim\limits_{x \to 2} \dfrac{x^2 - 4}{x - 2} = \lim\limits_{x \to 2} \dfrac{(x-2)(x+2)}{x - 2}$ [Note that $x^2 - 4 = x^2 - 2^2$ and recall that $a^2 - b^2 = (a + b)(a - b)$]

Cancelling out $(x - 2)$ gives:

$\lim\limits_{x \to 2} (x + 2)$

We now substitute 2 for x to obtain our final answer as follows:

$\lim\limits_{x \to 2} (x + 2) = 2 + 2$

$= 4$

Therefore, $\lim\limits_{x \to 2} \dfrac{x^2 - 4}{x - 2} = 4$

5. Evaluate $\lim\limits_{x \to 0} (2x - 5)(x + 2)(3x - 1)$

Solution

$\lim\limits_{x \to 0} (2x - 5)(x + 2)(3x - 1)$

Substituting 0 for x in the expression gives:

$\lim\limits_{x \to 0} (2x - 5)(x + 2)(3x - 1) = [2(0) - 5][(0) + 2][3(0) - 1]$

$= (-5)(2)(-1)$

$= 10$

6. Evaluate $\lim\limits_{x \to \infty} \dfrac{x^3 + 2x^2 + 5x + 3}{x^3 + x^2 + 3x + 7}$

Solution

In this case of x approaching infinity, we evaluate it by first dividing each term in the numerator and denominator by the variable having the highest exponent (power). Hence we divide each term by x^3. This is done as follows:

$\lim\limits_{x \to \infty} \dfrac{x^3 + 2x^2 + 5x + 3}{x^3 + x^2 + 3x + 7} = \lim\limits_{x \to \infty} \dfrac{\dfrac{x^3}{x^3} + \dfrac{2x^2}{x^3} + \dfrac{5x}{x^3} + \dfrac{3}{x^3}}{\dfrac{x^3}{x^3} + \dfrac{x^2}{x^3} + \dfrac{3x}{x^3} + \dfrac{7}{x^3}}$

$= \lim\limits_{x \to \infty} \dfrac{1 + \dfrac{2}{x} + \dfrac{5}{x^2} + \dfrac{3}{x^3}}{1 + \dfrac{1}{x} + \dfrac{3}{x^2} + \dfrac{7}{x^3}}$

$= \dfrac{1 + \dfrac{2}{\infty} + \dfrac{5}{\infty^2} + \dfrac{3}{\infty^3}}{1 + \dfrac{1}{\infty} + \dfrac{3}{\infty^2} + \dfrac{7}{\infty^3}}$

$$= \frac{1+0+0+0}{1+0+0+0}$$ (Note that a number divided by ∞ gives 0)

$$= \frac{1}{1}$$

$$= 1$$

7. Evaluate $\lim\limits_{x \to \infty} \dfrac{2x^4 - 3x^2 + 8x - 1}{x^4 - 5x + 3}$

Solution

The variable in its highest exponent (power) is x^4. Hence we divide each term by x^4 as follows:

$$\lim_{x \to \infty} \frac{2x^4 - 3x^2 + 8x - 1}{x^4 - 5x + 3} = \lim_{x \to \infty} \frac{\dfrac{2x^4}{x^4} - \dfrac{3x^2}{x^4} + \dfrac{8x}{x^4} - \dfrac{1}{x^4}}{\dfrac{x^4}{x^4} - \dfrac{5x}{x^4} + \dfrac{3}{x^4}}$$

$$= \lim_{x \to \infty} \frac{2 - \dfrac{3}{x^2} + \dfrac{8}{x^3} - \dfrac{1}{x^4}}{1 - \dfrac{5}{x^3} + \dfrac{3}{x^4}}$$

$$= \frac{2 - \dfrac{3}{\infty} + \dfrac{8}{\infty} - \dfrac{1}{\infty}}{1 - \dfrac{5}{\infty} + \dfrac{3}{\infty}}$$

$$= \frac{2 - 0 + 0 - 0}{1 - 0 + 0}$$ (Note that a number divided by ∞ gives 0)

$$= \frac{2}{1}$$

$$= 2$$

8. Evaluate $\lim\limits_{x \to 4} \dfrac{2x^2 - 7x - 4}{x^2 - 3x - 4}$

Solution

If 4 is substituted for x in the function above, it will give $\dfrac{0}{0}$ which is an indeterminate value.

Hence we factorize the expression and simplify as follows:

$$\lim_{x \to 4} \frac{2x^2 - 7x - 4}{x^2 - 3x - 4} = \lim_{x \to 4} \frac{(2x+1)(x-4)}{(x+1)(x-4)}$$

Cancelling out $(x - 4)$ gives:

$$\lim_{x \to 4} \frac{2x + 1}{x + 1}$$

We now substitute 4 for x to obtain our final answer as follows:

$$\lim_{x \to 4} \frac{2x + 1}{x + 1} = \frac{2(4) + 1}{(4) + 1}$$

$$= \frac{8+1}{4+1}$$

$$= \frac{9}{5}$$

$$= 1\frac{4}{5}$$

Therefore, $\lim\limits_{x \to 4} \dfrac{2x^2 - 7x - 4}{x^2 - 3x - 4} = 1\dfrac{4}{5}$

9. Evaluate $\lim\limits_{x \to 3} \dfrac{x^2 - 9}{x^2 + 6x + 9}$

<u>Solution</u>

Substituting 3 for x in the function can give us the final answer as follows:

$$\lim\limits_{x \to 3} \frac{x^2 - 9}{x^2 + 6x + 9} = \frac{3^2 - 9}{3^2 + 6(3) + 9}$$

$$= \frac{9 - 9}{9 + 18 + 9}$$

$$= \frac{0}{36}$$

$$= 0$$

Hence, $\lim\limits_{x \to 3} \dfrac{x^2 - 9}{x^2 + 6x + 9} = 0$

Note that even if this expression had been factorized and simplified, the final answer would still give zero.

10. Determine the limiting value of $\dfrac{x + 2}{x^2 + 6x + 8}$ as x tends to -2 .

<u>Solution</u>

The question can also be written as $\lim\limits_{x \to -2} \dfrac{x + 2}{x^2 + 6x + 8}$

If -2 is substituted for x in the function above, it will give $\dfrac{0}{0}$. Hence we factorize the expression and simplify as follows:

$$\lim\limits_{x \to -2} \frac{x + 2}{x^2 + 6x + 8} = \lim\limits_{x \to -2} \frac{x + 2}{(x + 2)(x + 4)}$$

Cancelling out $(x + 2)$ gives:

$$\lim\limits_{x \to -2} \frac{1}{x + 4}$$

We now substitute -2 for x as follows:

$$\lim\limits_{x \to -2} \frac{1}{x + 4} = \frac{1}{-2 + 4}$$

$$= \frac{1}{2}$$

11. Find the limit of $\dfrac{x^3 - 125}{x - 5}$ as $x \to 5$

Solution

The question can also be written as $\displaystyle\lim_{x \to 5} \dfrac{x^3 - 125}{x - 5}$

We factorize the expression and simplify as follows:

$$\lim_{x \to 5} \dfrac{x^3 - 125}{x - 5} = \lim_{x \to 5} \dfrac{x^3 - 5^3}{x - 5}$$

$$= \lim_{x \to 5} \dfrac{(x - 5)(x^2 + 5x + 5^2)}{x - 5} \quad \text{[Recall the identity } a^3 - b^3 = (a - b)(a^2 + ab + b^2)]$$

Cancelling out $(x - 5)$ gives:

$$\lim_{x \to 5} (x^2 + 5x + 5^2)$$

We now substitute 5 for x as follows:

$$\lim_{x \to 5} (x^2 + 5x + 5^2) = (5^2 + 5(5) + 25)$$

$$= 25 + 25 + 25$$

$$= 75$$

12. Find the limit of $\dfrac{x^3 + 64}{x + 4}$ as $x \to -4$

Solution

The question can also be written as $\displaystyle\lim_{x \to -4} \dfrac{x^3 + 4^3}{x + 4}$

We factorize the expression and simplify as follows:

$$\lim_{x \to -4} \dfrac{x^3 + 64}{x + 4} = \lim_{x \to -4} \dfrac{x^3 + 4^3}{x + 4}$$

$$= \lim_{x \to -4} \dfrac{(x + 4)(x^2 - 4x + 4^2)}{x + 4} \quad \text{[Recall the identity } a^3 + b^3 = (a + b)(a^2 - ab + b^2)]$$

Cancelling out $(x + 4)$ gives:

$$\lim_{x \to -4} (x^2 - 4x + 4^2)$$

We now substitute -4 for x as follows:

$$\lim_{x \to -4} (x^2 - 4x + 4^2) = (-4^2 - 4(-4) + 16)$$

$$= 16 + 16 + 16$$

$$= 48$$

13. Evaluate $\displaystyle\lim_{x \to 9} \dfrac{3 - \sqrt{x}}{9 - x}$

Solution

If 9 is substituted for x in the expression above, it will give $\dfrac{0}{0}$ which is not the desired answer.

Hence a way to go around this problem is to multiply the top and bottom by the conjugate of the surd at the top. The conjugate of $3 - \sqrt{x}$ is $3 + \sqrt{x}$ (only a difference in their middle signs).

Hence we multiply top and bottom by $3 + \sqrt{x}$ and simplify as follows:

$$\lim_{x \to 9} \frac{3 - \sqrt{x}}{9 - x} = \lim_{x \to 9} \frac{(3 - \sqrt{x})(3 + \sqrt{x})}{(9 - x)(3 + \sqrt{x})}$$

$$= \lim_{x \to 9} \frac{3^2 - (\sqrt{x})^2}{(9 - x)(3 + \sqrt{x})} \quad \text{(Note that the top was simplified by using } (a - b)(a + b) = a^2 - b^2)$$

$$= \lim_{x \to 9} \frac{9 - x}{(9 - x)(3 + \sqrt{x})}$$

$$= \lim_{x \to 9} \frac{1}{(3 + \sqrt{x})} \quad \text{(Since } 9 - x \text{ cancels out)}$$

We now substitute 9 for x as follows:

$$\lim_{x \to 9} \frac{1}{(3 + \sqrt{x})} = \frac{1}{(3 + \sqrt{9})}$$

$$= \frac{1}{3 + 3}$$

$$= \frac{1}{6}$$

14. Evaluate $\lim_{x \to 1} \dfrac{x^2 - 1}{x^5 - 1}$

Solution

If 2 is substituted for x in the expression above, it will give $\dfrac{0}{0}$ which is not a good answer. Hence a way to go around this problem is to divide the top and bottom by $x - 1$ (i.e. the difference between the variable and 1). This is done as follows:

$$\lim_{x \to 1} \frac{x^2 - 1}{x^5 - 1} = \lim_{x \to 1} \frac{\dfrac{x^2 - 1}{x - 1}}{\dfrac{x^5 - 1}{x - 1}} \quad \text{(Note that this has not changed the original fraction)}$$

Note that $\lim_{x \to a} \dfrac{x^n - a^n}{x - a} = na^{n-1}$. With this rule we now simplify the expression above as follows:

$$\lim_{x \to 1} \frac{\dfrac{x^2 - 1}{x - 1}}{\dfrac{x^5 - 1}{x - 1}} = \frac{2 \times 1^{(2-1)}}{5 \times 1^{(5-1)}}$$

$$= \frac{2 \times 1^1}{5 \times 1^5}$$

$$= \frac{2}{5}$$

Continuity of a function

A function is said to be continuous if the three conditions below are satisfied.

1. f(a) exists

2. $\lim\limits_{x \to a} f(x)$ exists

3. $\lim\limits_{x \to a} f(x) = f(a)$ (i.e. if the values of conditions 1 and 2 above are equal)

Examples

1. Determine if the function $f(x) = 2x^2 - 8x + 5$ is continuous at the point $x = -1$

Solution

When $x = -1$, the f(−1) is obtained as follows:

$$f(x) = 2x^2 - 8x + 5$$
$$f(-1) = 2(-1)^2 - 8(-1) + 5$$
$$= 2 + 8 + 5$$
$$= 15$$

Since f(−1) = 15, it means that f(−1) exists.

The next step is to find the value of $\lim\limits_{x \to -1} f(x)$ as follows:

$$\lim\limits_{x \to -1} 2x^2 - 8x + 5 = 2(-1)^2 - 8(-1) + 5$$
$$= 2 + 8 + 5$$
$$= 15$$

Hence $f(-1) = \lim\limits_{x \to -1} f(x) = 15$

Therefore, the function is continuous.

2. Determine if the function $f(x) = \dfrac{2x + 1}{x^2 - 3x + 7}$ is continuous at $x = 2$

Solution

When $x = 2$, the f(2) is obtained as follows:

$$f(x) = \dfrac{2x + 1}{x^2 - 3x + 7}$$
$$f(2) = \dfrac{2(2) + 1}{2^2 - 3(2) + 7}$$
$$= \dfrac{4 + 1}{4 - 6 + 7}$$
$$= \dfrac{5}{5}$$
$$= 1$$

Since f(2) = 1, it means that f(2) exists.

The next step is to find the value of $\lim\limits_{x \to 2} f(x)$ as follows:

$$\lim\limits_{x \to 2} \dfrac{2x + 1}{x^2 - 3x + 7} = \dfrac{2(2) + 1}{2^2 - 3(2) + 7}$$
$$= \dfrac{4 + 1}{4 - 6 + 7}$$
$$= \dfrac{5}{5}$$

$$= 1$$

Hence $f(2) = \lim_{x \to 2} f(x) = 1$

Therefore, the function is continuous.

3. Determine if the function $f(x) = \dfrac{2x^2 - 8}{x + 4}$ is continuous at $x = -4$.

When $x = -4$, the $f(-4)$ is obtained as follows:

$$f(x) = \frac{2x^2 - 8}{x + 4}$$

$$f(-4) = \frac{2(-4^2) - 8}{-4 + 4}$$

$$= \frac{2(16) - 8}{0}$$

$$= \frac{24}{0} \quad \text{(Undefined)}$$

Hence $f(-4)$ is undefined, and does not exist. This shows that the function is not continuous. It is discontinuous as $x = -4$.

4. Determine if the function $f(x) = \dfrac{x^2 - 9}{x - 3}$ is continuous at $x = 3$.

When $x = 3$, the $f(3)$ is obtained as follows:

$$f(x) = \frac{x^2 - 9}{x - 3}$$

$$f(3) = \frac{3^2 - 9}{3 - 3}$$

$$= \frac{0}{0} \quad \text{(This has no value)}$$

Hence $f(3)$ does not exist.

Let us determine the value of $\lim_{x \to 3} f(x)$ as follows:

$$\lim_{x \to 3} \frac{x^2 - 9}{x - 3} = \lim_{x \to 3} \frac{(x - 3)(x + 3)}{x - 3}$$

$$= \lim_{x \to 3} (x + 3) \quad \text{(Since } x - 3 \text{ cancels out)}$$

$$= 3 + 3 \quad \text{(When 3 is substituted for } x\text{)}$$

$$= 6$$

Hence, $f(3)$ has no value while $\lim_{x \to 3} f(x) = 6$

This shows that $f(3) \neq \lim_{x \to 3} f(x)$

Therefore the function discontinuous.

5. Test for the continuity of the function: $f(x) = \begin{cases} -5x + 7 & \text{if } x > 1 \\ x^2 + 2 & \text{if } x \leq 1 \end{cases}$

Solution

$$f(x) = \begin{cases} -5x + 7 & \text{if } x > 1 \\ x^2 + 2 & \text{if } x \leq 1 \end{cases}$$

169

When $x = 1$, we use the expression $x^2 + 2$ to evaluate f(1). Note that $x \leq 1$ means $x < 1$ or $x = 1$.

Hence, $f(x) = x^2 + 2$ (for $x = 1$)

$$f(1) = 1^2 + 2$$
$$= 3$$

Since f(1) = 3, it means that f(1) exist.

Let us now determine the limiting value of f(x).

If x approaches 1 from the left, which is $x \leq 1$ (left hand elbow) we write it as $\lim_{x \to 1^-} f(x)$, and we pick the expression given by:

$$f(x) = x^2 + 2$$

Hence, $\lim_{x \to 1} f(x) = 1^2 + 3$

$$= 3$$

If x approaches 1 from the right, which is $x > 1$ (right hand elbow), we write it as $\lim_{x \to 1^+} f(x)$, and we pick the expression given by:

$$f(x) = -5x + 7$$

Hence, $\lim_{x \to 1} f(x) = -5(1) + 7$

$$= 2$$

Hence, $\lim_{x \to 1^-} f(x) = 3$ while $\lim_{x \to 1^+} f(x) = 2$. They are not equal. This shows that $\lim_{x \to 1} f(x)$ does not exist. Therefore, the function is not continuous.

6. Determine the continuity of the function: $f(x) = \begin{cases} x^2 & \text{if } x \geq -2 \\ x+6 & \text{if } x < -2 \end{cases}$

<u>Solution</u>

$$f(x) = \begin{cases} x^2 & \text{if } x \geq -2 \\ x+6 & \text{if } x < -2 \end{cases}$$

A direct way of determining the continuity of functions such as these is to find the limits of the two expressions in the function as x tends to the values given in the question.

Therefore, when x tends to -2 in the function $f(x) = x^2$, we have:

$$f(x) = x^2$$

Hence, $\lim_{x \to -2} f(x) = (-2)^2$

$$= 4$$

Also, when x tends to -2 in the function $f(x) = x + 6$, we have:

$$f(x) = x + 6$$

Hence, $\lim_{x \to -2} f(x) = -2 + 6$

$$= 4$$

Hence, $\lim_{x \to -2} x^2 = \lim_{x \to -2} x + 6 = 4$. Their limits give the same value. This shows that $\lim_{x \to -2} f(x)$ exist. Therefore, the function is continuous.

7. Determine if the function below is continuous or not.

$$f(x) = \begin{cases} \dfrac{x^2 - 16}{x + 4} & \text{if } x \geq -4 \\ \dfrac{x + 4}{2x - 3} & \text{if } x < -4 \end{cases}$$

<u>Solution</u>

As x tends to -4 in the function $f(x) = \dfrac{x^2 - 16}{x + 4}$ we simplify as follows:

$$f(x) = \frac{(x + 4)(x - 4)}{x + 4}$$

Hence, $\underset{x \to -4}{\lim} f(x) = (x - 4)$ (Note that $(x + 4)$ cancels out)

$$= -4 - 4$$
$$= -8$$

Also, when x tends to -4 in the function $f(x) = \dfrac{x + 4}{2x - 3}$, we have:

$$\underset{x \to -4}{\lim} f(x) = \frac{-4 + 4}{2(-4) - 3}$$
$$= \frac{-4 + 4}{-8 - 3}$$
$$= \frac{0}{-11}$$
$$= 0$$

Their limits give different values. This shows that $\underset{x \to -4}{\lim} f(x)$ does not exist. Therefore, the function is not continuous.

8. If $f(x) = \begin{cases} \dfrac{x^2 - 4}{x - 2} \\ 4 \ \text{ if } x = 2 \end{cases}$ test for the continuity of the function at $x = 2$.

<u>Solution</u>

$$f(x) = \begin{cases} \dfrac{x^2 - 4}{x - 2} \\ 4 \ \text{ if } x = 2 \end{cases}$$

Let us find the limits of the two expressions in the function as x tends 2.

From the first expression, as x tends to 2 we have:

$$f(x) = \frac{x^2 - 4}{x - 2}$$

Hence, $\underset{x \to 2}{\lim} f(x) = \dfrac{(x - 2)(x + 2)}{x - 2}$ (When the numerator is factorized)

$\underset{x \to 2}{\lim} f(x) = (x + 2)$ $(x - 2$ has cancelled out)

$\underset{x \to 2}{\lim} f(x) = 2 + 2$

$$= 4$$

Also, when x tends to 2 in the function $f(x) = 4$, we have:

$$f(x) = 4$$

Hence, $\lim\limits_{x \to 2} 4 = 4$ (Since $\lim\limits_{x \to a} c = c$)

The question also tells us that when $x = 2$, f(x) = 4

Hence, $\lim\limits_{x \to 2} 4 = \lim\limits_{x \to 2} \dfrac{x^2 - 4}{x - 2} = 4$. Their limits give the same value of 4. This shows that f(x) exist,

and that f(2) = $\lim\limits_{x \to 2}$ f(x). Therefore, the function is continuous.

Examples on Limits of Trigonometric Functions

1. Evaluate $\lim\limits_{x \to 0} \dfrac{\sin 2x}{5x}$

Solution

$\lim\limits_{x \to 0} \dfrac{\sin 2x}{5x}$

Substituting zero directly into the expression above will give $\dfrac{0}{0}$. Hence we have to apply the

appropriate rule of limit. Let us first make an adjustment to the expression as follows.

Multiply the expression by $\dfrac{2x}{5x}$ and change the denominator to $2x$ just like the numerator, as

follows:

$$\lim\limits_{x \to 0} \dfrac{\sin 2x}{5x} = \lim\limits_{x \to 0} \left(\dfrac{\sin 2x}{2x} \times \dfrac{2x}{5x}\right)$$

Note that when $2x$ cancels out, the original expression remains the same.

Recall that $\lim\limits_{x \to 0} \dfrac{\sin ax}{ax} = 1$

Hence, $\lim\limits_{x \to 0} \dfrac{\sin 2x}{2x} = 1$

Substituting 1 for $\lim\limits_{x \to 0} \dfrac{\sin 2x}{2x}$ in the simplification given above to gives:

$$\lim\limits_{x \to 0} \left(\dfrac{\sin 2x}{2x} \times \dfrac{2x}{5x}\right) = \lim\limits_{x \to 0} \dfrac{\sin 2x}{2x} \times \lim\limits_{x \to 0} \dfrac{2x}{5x}$$

$$= 1 \times \lim\limits_{x \to 0} \dfrac{2}{5} \quad \text{(x cancels out)}$$

$$= 1 \times \dfrac{2}{5} \quad \text{(Since } \lim\limits_{x \to a} c = c\text{)}$$

$$= \dfrac{2}{5}$$

2. 1. Evaluate $\lim\limits_{x \to 0} \dfrac{\sin 5x}{7x}$

Solution

$\lim\limits_{x \to 0} \dfrac{\sin 5x}{7x}$

Multiply the expression by $\dfrac{5x}{7x}$ and change the denominator to $5x$ just like the numerator. This

gives:

172

$$\lim_{x \to 0} \frac{\sin 5x}{7x} = \lim_{x \to 0} \left(\frac{\sin 5x}{5x} \times \frac{5x}{7x} \right)$$

Hence, $\lim_{x \to 0} \frac{\sin 5x}{5x} = 1$

Substituting 1 for $\lim_{x \to 0} \frac{\sin 5x}{5x}$ in the simplification given above to gives:

$$\lim_{x \to 0} \left(\frac{\sin 5x}{5x} \times \frac{5x}{7x} \right) = \lim_{x \to 0} \frac{\sin 5x}{5x} \times \lim_{x \to 0} \frac{5x}{7x}$$

$$= 1 \times \lim_{x \to 0} \frac{5}{7} \quad (x \text{ cancels out})$$

$$= 1 \times \frac{5}{7} \quad (\text{Since } \lim_{x \to a} c = c)$$

$$= \frac{5}{7}$$

The two examples above show that $\lim_{x \to 0} \frac{\sin a\, x}{x} = a$ or $\lim_{x \to 0} \frac{\sin a\, x}{bx} = \frac{a}{b}$. This rule also applies to tangent as $\lim_{x \to 0} \frac{\tan a\, x}{x} = a$ or $\lim_{x \to 0} \frac{\tan a\, x}{bx} = \frac{a}{b}$.

3. Find the value of $\lim_{x \to \frac{\pi}{2}} \frac{\cos 2x}{\sin 3x}$

Solution

$$\lim_{x \to \frac{\pi}{2}} \frac{\cos 2x}{\sin 3x}$$

We can simply substitute in the value of x and obtain our answer.

$$\lim_{x \to \frac{\pi}{2}} \frac{\cos 2x}{\sin 3x} = \frac{\cos 2\left(\frac{\pi}{2}\right)}{\sin 3\left(\frac{\pi}{2}\right)}$$

$$= \frac{\cos \pi}{\sin \frac{3\pi}{2}}$$

$$= \frac{\cos 180}{\sin 270} \quad (\text{Note that } \pi \text{ radians} = 180°)$$

$$= \frac{-1}{-1}$$

$$= 1$$

This problem can also be solved by using the angles directly in radians, but I prefer to work in degrees. Note that cos π in radians = −1, and $\sin \frac{3\pi}{2}$ in radians = −1 as obtained above.

4. Evaluate $\lim_{x \to 0} \frac{\cos x}{\sin x - 3}$

Solution

$$\lim_{x \to 0} \frac{\cos x}{\sin x - 3} = \frac{\cos 0}{\sin 0 - 3}$$

$$= \frac{1}{0-3}$$

$$= -\frac{1}{3}$$

5. Evaluate $\lim\limits_{x \to 0} \dfrac{\sin 3x}{\sin 2x}$

Solution

Multiply by $\dfrac{x}{\sin 2x}$ and change the denominator to x as follows.

$$\lim\limits_{x \to 0} \frac{\sin 3x}{\sin 2x} = \lim\limits_{x \to 0} \left(\frac{\sin 3x}{x} \; \text{X} \; \frac{x}{\sin 2x} \right) \qquad \text{(When } x \text{ cancels out it gives the original expression)}$$

$$= \lim\limits_{x \to 0} \left(\frac{\sin 3x}{x} \; \text{X} \; \left(\frac{\sin 2x}{x} \right)^{-1} \right) \qquad \text{(Note that the inverse of } \frac{x}{\sin 2x} \text{ has been taken)}$$

$$= \lim\limits_{x \to 0} \frac{\sin 3x}{x} \; \text{X} \; \left(\lim\limits_{x \to 0} \frac{\sin 2x}{x} \right)^{-1}$$

$$= 3 \text{ x } 2^{-1} \qquad \text{(Note that } \lim\limits_{x \to 0} \frac{\sin 3x}{x} = 3 \text{ and } \lim\limits_{x \to 0} \frac{\sin 2x}{x} = 2)$$

$$= 3 \text{ x } \frac{1}{2}$$

$$= \frac{3}{2}$$

This example shows that: $\lim\limits_{x \to 0} \dfrac{\sin a \, x}{\sin b \, x} = \dfrac{a}{b}$

6. Evaluate $\lim\limits_{x \to 0} \dfrac{\sin 2x}{\sin 5x}$

Solution

$$\lim\limits_{x \to 0} \frac{\sin 2x}{\sin 5x}$$

From the rule established in example 5 above, we can see that the solution to this problem is $\dfrac{2}{5}$.

Hence, $\lim\limits_{x \to 0} \dfrac{\sin 2x}{\sin 5x} = \dfrac{2}{5} \qquad$ (Since $\lim\limits_{x \to 0} \dfrac{\sin a \, x}{\sin b \, x} = \dfrac{a}{b}$)

Exercise 12

1. Evaluate $\lim\limits_{x \to 0} 2x^3 - 4x^2 + x + 9$

2. Evaluate $\lim\limits_{x \to 0} \dfrac{5x^2 + x - 8}{x^2 - 2x + 5}$

3. Evaluate $\lim\limits_{x \to 5} \dfrac{x^2 - x - 20}{x - 5}$

4. Evaluate $\lim\limits_{x \to 7} \dfrac{x^2 - 49}{x - 7}$

5. Evaluate $\lim\limits_{x \to 0} (5x - 1)(3x + 7)(3x + 2)$

6. Evaluate $\lim\limits_{x \to \infty} \dfrac{2x^3 + x^2 - 9x - 5}{5x^3 + 2x^2 - 7x + 2}$

7. Evaluate $\lim\limits_{x \to \infty} \dfrac{x^4 - 5x^2 - 2x - 6}{3x^4 + 2x + 3}$

8. Evaluate $\lim\limits_{x \to 4} \dfrac{x^2 - 25}{x^2 + 2x - 15}$

9. Evaluate $\lim\limits_{x \to 8} \dfrac{2x^2 - 17x + 8}{8 - x}$

10. Evaluate $\lim\limits_{m \to 0} \dfrac{(5+m)^2 - 25}{m}$

11. Determine the limiting value of $\dfrac{\sqrt{x} - 2}{x - 4}$ as x tends to 4 .

12. Find the limit of $\dfrac{x^3 - 27}{x - 3}$ as $x \to 3$

13. Find the limit of $\dfrac{x^3 + 8}{x + 2}$ as $x \to -2$

14. Evaluate $\lim\limits_{x \to 25} \dfrac{5 - \sqrt{x}}{25 - x}$

15. Evaluate $\lim\limits_{x \to 1} \dfrac{x^3 - 1}{x^7 - 1}$

16. Determine if the function $f(x) = x^2 + 3x - 1$ is continuous at the point $x = 2$

17. Determine if the function $f(x) = \dfrac{5x - 2}{3x^2 - x - 2}$ is continuous at $x = -1$

18. Determine if the function $f(x) = \dfrac{2x^2 - 18}{x - 3}$ is continuous at $x = 3$.

19. Determine if the function $f(x) = \dfrac{x^3 + 64}{x + 4}$ is continuous at $x = -4$.

20. Test for the continuity of the function: $f(x) = \begin{cases} -x+5 & \text{if } x > 2 \\ x^3 + 10 & \text{if } x \le 2 \end{cases}$

21. Determine the continuity of the function: $f(x) = \begin{cases} 3x^2 & \text{if } x \ge -3 \\ x+30 & \text{if } x < -3 \end{cases}$

22. Determine if the function below is continuous or not.

$$f(x) = \begin{cases} \dfrac{x^2 - 25}{x + 5} & \text{if } x \ge -5 \\ \dfrac{x + 5}{2x + 10} & \text{if } x < -5 \end{cases}$$

23. If $f(x) = \begin{cases} \dfrac{x^3 - 27}{x - 3} \\ 1 \text{ if } x = 3 \end{cases}$ test for the continuity of the function at $x = 3$.

24. If $f(x) = \begin{cases} 2 - 3x & x < 1 \\ x^3 + 4 & x \ge 1 \end{cases}$ evaluate the following limits, if they exist.

(a) $\lim\limits_{x \to -2} f(x)$

(b) $\lim\limits_{x \to 1} f(x)$

25. Evaluate $\lim\limits_{x \to 2} (5 + |x - 2|)$ if it exists.

26. Evaluate $\lim\limits_{x \to 0} \dfrac{\sin 7x}{2x}$

27. Evaluate $\lim\limits_{x \to 0} \dfrac{\sin x}{3x}$

28. Find the value of $\lim\limits_{x \to \pi} \dfrac{\cos 2x}{\sin \frac{x}{2}}$

29. Evaluate $\lim\limits_{x \to 0} \dfrac{\cos 5x}{\sin 3x - 2}$

30. Evaluate $\lim\limits_{x \to 0} \dfrac{\sin 9x}{\sin 4x}$

CHAPTER 13
DIFFERENTIATION FROM FIRST PRINCIPLE

If y = f(x), then the gradient function of y or f(x) is given by:

$$g(x) = \frac{f(x + \Delta x) - f(x)}{\Delta x}$$

where Δx is the increment in x.

Applying limits has shown that as Δx tends to zero,

$$f'(x) = \lim_{\Delta x \to 0} \frac{f(x + \Delta x) - f(x)}{\Delta x}$$

$$\frac{dy}{dx} = \lim_{\Delta x \to 0} \frac{f(x + \Delta x) - f(x)}{\Delta x}$$

$\dfrac{dy}{dx}$ reads dee y dee x.

Note that f'(x) and $\dfrac{dy}{dx}$ can be interpreted as instantaneous rate of change of y with respect to x.

The technique of finding the derivative of a function by considering the limiting value is called differentiation from first principle.

Examples

1. Find the derivative of x from first principle.

Solution

Let, y = x

Increasing x by Δx will give an increase in y by Δy. This gives:

y + Δy = x + Δx

Δy = x + Δx − y

Δy = x + Δx − x (substitute x for y since y = x as given above from the question)

Δy = Δx (x cancels out)

Dividing both sides by Δx in order to get the derivative gives:

$$\frac{\Delta y}{\Delta x} = \frac{\Delta x}{\Delta x}$$

$$\frac{\Delta y}{\Delta x} = 1$$

Taking limits as Δx tends to zero gives:

$\displaystyle\lim_{\Delta x \to 0} \frac{\Delta y}{\Delta x} = 1$ [Recall that $\displaystyle\lim_{x \to 0} c = c$, hence $\displaystyle\lim_{\Delta x \to 0} f(\Delta x) = 1$ gives 1, since $\dfrac{\Delta y}{\Delta x}$ is regarded as f(Δx)]

We now write $\displaystyle\lim_{\Delta x \to 0} \frac{\Delta y}{\Delta x}$ as $\dfrac{dy}{dx}$. This gives:

$$\frac{dy}{dx} = 1$$

2. Find from first principle, the derivative of $2x^3$.

<u>Solution</u>

Let, $y = 2x^3$

Increasing x by Δx will give an increase in y by Δy. This gives:

$$y + \Delta y = 2(x + \Delta x)^3$$

$$y + \Delta y = 2[x^3 + 3x^2\Delta x + 3x(\Delta x)^2 + (\Delta x)^3]$$

$$= 2x^3 + 6x^2\Delta x + 6x(\Delta x)^2 + 2(\Delta x)^3$$

$$\Delta y = 2x^3 + 6x^2\Delta x + 6x(\Delta x)^2 + 2(\Delta x)^3 - y$$

$$= 2x^3 + 6x^2\Delta x + 6x(\Delta x)^2 + 2(\Delta x)^3 - 2x^3 \quad \text{(Since } y = 2x^3\text{)}$$

$$\Delta y = 6x^2\Delta x + 6x(\Delta x)^2 + 2(\Delta x)^3 \quad \text{(Since } 2x^3 \text{ cancels out)}$$

Dividing both sides by Δx in order to get the derivative gives:

$$\frac{\Delta y}{\Delta x} = \frac{6x^2\Delta x}{\Delta x} + \frac{6x(\Delta x)^2}{\Delta x} + \frac{2(\Delta x)^3}{\Delta x}$$

$$\frac{\Delta y}{\Delta x} = 6x^2 + 6x\Delta x + 2(\Delta x)^2$$

Taking limits as Δx tends to zero gives:

$$\lim_{\Delta x \to 0} \frac{\Delta y}{\Delta x} = 6x^2 + 6x(0) + 2(0)^2 \quad \text{(Note that zero replaces } \Delta x\text{)}$$

$$\lim_{\Delta x \to 0} \frac{\Delta y}{\Delta x} = 6x^2$$

We now write $\lim_{\Delta x \to 0} \frac{\Delta y}{\Delta x}$ as $\frac{dy}{dx}$. This gives:

$$\frac{dy}{dx} = 6x^2$$

3. Find the derivative of $f(x) = \dfrac{1}{x^2}$ from first principle.

<u>Solution</u>

$$f(x) = \Delta x$$

Increasing x by Δx gives:

$$f(x + \Delta x) = \frac{1}{(x + \Delta x)^2}$$

$$f(x + \Delta x) = \frac{1}{x^2 + 2x\Delta x + (\Delta x)^2}$$

Subtract f(x) from both sides of the equation

$$f(x + \Delta x) - f(x) = \frac{1}{x^2 + 2x\Delta x + (\Delta x)^2} - f(x)$$

$$f(x + \Delta x) - f(x) = \frac{1}{x^2 + 2x\Delta x + (\Delta x)^2} - \frac{1}{x^2} \quad \text{(Note that } f(x) = \frac{1}{x^2}\text{)}$$

$$= \frac{x^2 - (x^2 + 2x\Delta x + (\Delta x)^2)}{x^2(x^2 + 2x\Delta x + (\Delta x)^2)}$$

$$= \frac{x^2 - x^2 - 2x\Delta x - (\Delta x)^2}{x^2(x^2 + 2x\Delta x + (\Delta x)^2)}$$

$$= \frac{-2x\Delta x - (\Delta x)^2}{x^2(x^2 + 2x\Delta x + (\Delta x)^2)}$$

$$f(x + \Delta x) - f(x) = \frac{\Delta x(-2x - \Delta x)}{x^2(x^2 + 2x\Delta x + (\Delta x)^2)}$$

Dividing both sides by Δx gives:

$$\frac{f(x + \Delta x) - f(x)}{\Delta x} = \frac{-2x - \Delta x}{x^2(x^2 + 2x\Delta x + (\Delta x)^2)}$$ (Δx has cancelled out from the right hand side)

Taking limits as Δx tends to zero gives:

$$\lim_{\Delta x \to 0} \frac{f(x + \Delta x) - f(x)}{\Delta x} = \frac{-2x - 0}{x^2(x^2 + 2x(0) + (0)^2)}$$

Replacing $\lim_{\Delta x \to 0} \frac{f(x + \Delta x) - f(x)}{\Delta x}$ with f'(x) gives the derivative of f(x) as:

$$f'(x) = \frac{-2x}{x^2(x^2)}$$

$$f'(x) = \frac{-2}{x^3}$$

4. If $f(x) = 3x^2$

(a) write down and simplify the expression $\frac{f(x + h) - f(x)}{h}$ ($h \neq 0$)

(b) find $\lim_{\Delta x \to 0} \frac{f(x + h) - f(x)}{h}$

Solution

(a) Let, $f(x) = 3x^2$

Increasing x by h gives:

$$f(x + h) = 3(x + h)^2$$
$$= 3[x^2 + 2xh + h^2]$$
$$= 3x^2 + 6xh + 3h^2$$

Subtracting f(x) from both sides gives:

$$f(x + h) - f(x) = 3x^2 + 6xh + 3h^2 - 3x^2$$ (Note that $f(x) = 3x^2$ from the question)
$$= 6xh + 3h^2$$

Dividing both side by h gives:

$$\frac{f(x + h) - f(x)}{h} = \frac{6xh}{h} + \frac{3h^2}{h}$$

$$\frac{f(x + h) - f(x)}{h} = 6x + 3h$$

(b) $\frac{f(x + h) - f(x)}{h} = 6x + 3h$

Taking limits as h tends to zero gives:

$$\lim_{h \to 0} \frac{f(x + h) - f(x)}{h} = 6x + 3(0)$$

$$\lim_{h \to 0} \frac{f(x + h) - f(x)}{h} = 6x$$

5. If $f(x) = 4x^3 - 5$,

(a) evaluate $\dfrac{f(x+h) - f(x)}{h}$, where $h \neq 0$

(b) From your result in (5a) above, find the derivatives of $f(x)$ with respect to x.

Solution

(a) Let, $f(x) = 4x^3 - 5$

Increasing x by h gives:

$$f(x+h) = 4(x+h)^3 - 5$$
$$= 4[x^3 + 3x^2h + 3xh^2 + h^3] - 5$$
$$= 4x^3 + 12x^2h + 12xh^2 + 4h^3 - 5$$

Subtracting $f(x)$ from both sides gives:

$$f(x+h) - f(x) = 4x^3 + 12x^2h + 12xh^2 + 4h^3 - 5 - (4x^3 - 5) \quad (f(x) = 4x^3 - 5 \text{ from the question})$$
$$= 4x^3 + 12x^2h + 12xh^2 + 4h^3 - 5 - 4x^3 + 5$$
$$= 12x^2h + 12xh^2 + 4h^3$$

Dividing both side by h gives:

$$\dfrac{f(x+h) - f(x)}{h} = 12x^2 + 12xh + 4h^2$$

(b) $\dfrac{f(x+h) - f(x)}{h} = 12x^2 + 12xh + 4h^2$

Taking limits as h tends to zero gives:

$$\lim_{h \to 0} \dfrac{f(x+h) - f(x)}{h} = 12x^2 + 12x(0)\ 4(0)^2$$

$$\lim_{h \to 0} \dfrac{f(x+h) - f(x)}{h} = 12x^2$$

6. If $f(x) = \dfrac{x^2 + 1}{2x}$

(a) write down and simplify the expression $\dfrac{f(x + \Delta x) - f(\Delta x)}{\Delta x}$, where $\Delta x \neq 0$

(b) Find the derivatives of $y = \dfrac{x^2 + 1}{2x}$

Solution

$$f(x) = \dfrac{x^2 + 1}{2x}$$

$$f(x + \Delta x) = \dfrac{(x + \Delta x)^2 + 1}{2(x + \Delta x)}$$

$$f(x + \Delta x) = \dfrac{x^2 + 2x\Delta x + (\Delta x)^2 + 1}{2x + 2\Delta x}$$

Subtract $f(x)$ from both sides of the equation

$$f(x + \Delta x) - f(x) = \dfrac{x^2 + 2x\Delta x + (\Delta x)^2 + 1}{2x + 2\Delta x} - \dfrac{x^2 + 1}{2x}$$

180

$$f(x + \Delta x) - f(x) = \frac{2x[x^2 + 2x\Delta x + (\Delta x)^2 + 1] - (x^2 + 1)(2x + 2\Delta x)}{2x(2x + 2\Delta x)}$$

$$= \frac{2x^3 + 4x^2\Delta x + 2x(\Delta x)^2 + 2x - (2x^3 + 2x^2\Delta x + 2x + 2\Delta x)}{2x(2x + 2\Delta x)}$$

$$= \frac{2x^3 + 4x^2\Delta x + 2x(\Delta x)^2 + 2x - 2x^3 - 2x^2\Delta x - 2x - 2\Delta x)}{2x(2x + 2\Delta x)}$$

$$= \frac{2x^2\Delta x + 2x(\Delta x)^2 - 2\Delta x)}{2x(2x + 2\Delta x)}$$

$$f(x + \Delta x) - f(x) = \frac{\Delta x(2x^2 + 2x\Delta x - 2)}{2x(2x + 2\Delta x)} \qquad \text{(After factorizing the numerator)}$$

Dividing both sides by Δx gives:

$$\frac{f(x + \Delta x) - f(x)}{\Delta x} = \frac{2x^2 + 2x\Delta x - 2}{2x(2x + 2\Delta x)}$$

(b)Taking limits as Δx tends to zero gives:

$$\lim_{\Delta x \to 0} \frac{f(x + \Delta x) - f(x)}{\Delta x} = \frac{2x^2 + 2x(0) - 2}{2x(2x + 2(0))}$$

$$= \frac{2x^2 - 2}{2x(2x)}$$

$$= \frac{2(x^2 - 1)}{4x^2}$$

Replacing $\displaystyle\lim_{\Delta x \to 0} \frac{f(x + \Delta x) - f(x)}{\Delta x}$ with f'(x) gives the derivative of f(x) as:

$$f'(x) = \frac{x^2 - 1}{2x^2}$$

Separating into fractions gives:

$$f'(x) = \frac{x^2}{2x^2} - \frac{1}{2x^2}$$

$$f'(x) = \frac{1}{2} - \frac{1}{2x^2}$$

7. Find from first principle the derivatives with respect to x of $y = 5x^3 - x + 7$

Solution

$$y = 5x^3 - x + 7$$

$$y + \Delta y = 5(x + \Delta x)^3 - (x + \Delta x) + 7$$

$$y + \Delta y = 5[x^3 + 3x^2\Delta x + 3x(\Delta x)^2 + (\Delta x)^3] - x - \Delta x + 7$$

$$= 5x^3 + 15x^2\Delta x + 15x(\Delta x)^2 + 5(\Delta x)^3 - x - \Delta x + 7$$

$$\Delta y = 5x^3 + 15x^2\Delta x + 15x(\Delta x)^2 + 5(\Delta x)^3 - x - \Delta x + 7 - y$$

$$= 5x^3 + 15x^2\Delta x + 15x(\Delta x)^2 + 5(\Delta x)^3 - x - \Delta x + 7 - (5x^3 - x + 7) \quad \text{(Since } y = 5x^3 - x + 7)$$

$$\Delta y = 5x^3 + 15x^2\Delta x + 15x(\Delta x)^2 + 5(\Delta x)^3 - x - \Delta x + 7 - 5x^3 + x - 7$$

$$= 15x^2\Delta x + 15x(\Delta x)^2 + 5(\Delta x)^3 - \Delta x$$

Dividing both sides by Δx gives:

$$\frac{\Delta y}{\Delta x} = \frac{15x^2 \Delta x}{\Delta x} + \frac{15x(\Delta x)^2}{\Delta x} + \frac{5(\Delta x)^3}{\Delta x} - \frac{\Delta x}{\Delta x}$$

$$\frac{\Delta y}{\Delta x} = 15x^2 + 15x\Delta x + 5(\Delta x)^2 - 1$$

Taking limits as Δx tends to zero gives:

$$\lim_{\Delta x \to 0} \frac{\Delta y}{\Delta x} = 15x^2 + 15x(0) + 5(0)^2 - 1$$

$$\lim_{\Delta x \to 0} \frac{\Delta y}{\Delta x} = 15x^2 - 1$$

We now replace $\lim_{\Delta x \to 0} \frac{\Delta y}{\Delta x}$ with $\frac{dy}{dx}$. This gives:

$$\frac{dy}{dx} = 15x^2 - 1$$

8. Differentiate $3x - \dfrac{1}{2x^2}$ from first principle.

Solution

$$y = 3x - \frac{1}{x^2}$$

$$y + \Delta y = 3(x + \Delta x) - \frac{1}{2(x + \Delta x)^2}$$

$$= 3x + 3\Delta x - \frac{1}{2[x^2 + 2x\Delta x + (\Delta x)^2]}$$

Subtract y from both sides of the equation

$$\Delta y = 3x + 3\Delta x - \frac{1}{2x^2 + 4x\Delta x + 2(\Delta x)^2} - y$$

$$= 3x + 3\Delta x - \frac{1}{2x^2 + 4x\Delta x + 2(\Delta x)^2} - \left(3x - \frac{1}{2x^2}\right)$$

$$= 3x + 3\Delta x - \frac{1}{2x^2 + 4x\Delta x + 2(\Delta x)^2} - 3x + \frac{1}{2x^2}$$

$$= 3\Delta x - \frac{1}{2x^2 + 4x\Delta x + 2(\Delta x)^2} + \frac{1}{2x^2}$$

Combining them into one fraction gives:

$$\Delta y = \frac{3\Delta x[2x^2(2x^2 + 4x\Delta x + 2(\Delta x)^2] - 2x^2 + 2x^2 + 4x\Delta x + 2(\Delta x)^2}{2x^2[2x^2 + 4x\Delta x + 2(\Delta x)^2]}$$

$$= \frac{3\Delta x[4x^4 + 8x^3\Delta x + 4x^2(\Delta x)^2] + 4x\Delta x + 2(\Delta x)^2}{2x^2[2x^2 + 4x\Delta x + 2(\Delta x)^2]} \quad (-2x^2 + 2x^2 \text{ has cancelled out})$$

$$= \frac{12x^4\Delta x + 24x^3(\Delta x)^2 + 12x^2(\Delta x)^3 + 4x\Delta x + 2(\Delta x)^2}{2x^2[2x^2 + 4x\Delta x + 2(\Delta x)^2]}$$

Factorizing the numerator gives:

$$= \frac{\Delta x[12x^4 + 24x^3\Delta x + 12x^2(\Delta x)^2 + 4x + 2\Delta x]}{2x^2[2x^2 + 4x\Delta x + 2(\Delta x)^2]}$$

Dividing both sides by Δx gives:

$$\frac{\Delta y}{\Delta x} = \frac{12x^4 + 24x^3\Delta x + 12x^2(\Delta x)^2 + 4x + 2\Delta x}{2x^2[2x^2 + 4x\Delta x + 2(\Delta x)^2]}$$ (Δx that is outside the bracket cancels out)

Taking limits as Δx tends to zero gives:

$$\lim_{\Delta x \to 0} \frac{\Delta y}{\Delta x} = \frac{12x^4 + 24x^3(0) + 12x^2(0)^2 + 4x + 2(0)}{2x^2[2x^2 + 4x(0) + 2(0)^2]}$$

$$= \frac{12x^4 + 4x}{2x^2[2x^2]}$$

$$\lim_{\Delta x \to 0} \frac{\Delta y}{\Delta x} = \frac{12x^4 + 4x}{4x^4}$$

$$= \frac{12x^4}{4x^4} + \frac{4x}{4x^4}$$

$$= 3 + \frac{1}{x^3}$$

Replacing $\lim_{\Delta x \to 0} \frac{\Delta y}{\Delta x}$ with $\frac{dy}{dx}$ gives the derivative as:

$$\frac{dy}{dx} = 3 + \frac{1}{x^3}$$

Exercise 13

1. Find the derivative of $2x$ from first principle.

2. Find from first principle, the derivative of x^2.

3. Find the derivative of $f(x) = \frac{1}{x^3}$ from first principle.

4. If $f(x) = 5x^2$

(a) write down and simplify the expression $\frac{f(x+h) - f(x)}{h}$ $(h \neq 0)$

(b) find $\lim_{h \to 0} \frac{f(x+h) - f(x)}{h}$

5. If $f(x) = 9x^3$,

(a) evaluate $\frac{f(x+h) - f(x)}{h}$, where $h \neq 0$

(b) From your result above, find the derivatives of $f(x)$ with respect to x.

6. If $f(x) = \frac{x^3 - 2}{x}$

(a) write down and simplify the expression $\frac{f(x + \Delta x) - f(\Delta x)}{\Delta x}$, where $\Delta x \neq 0$

(b) Find the derivatives of $y = \frac{x^3 - 2}{x}$

7. Find from first principle the derivatives with respect to x of $y = 3x^2 - 10x$

8. Differentiate $x + \dfrac{3}{x}$ from first principle.

9. From first principle, find the derivatives with respect to x of $y = 5x - 3x^2$

10. Differentiate $2x + \dfrac{x}{5}$ from first principle.

CHAPTER 14
GENERAL RULE OF DIFFERENTIATION AND COMPOSITE FUNCTIONS

The general rule for the derivative/differentiation of a function is as given below.

If $y = x^n$

then, $\dfrac{dy}{dx} = nx^{n-1}$

The rule for differentiating a composite function (or function of a function) is given as follows:

If $y = (2x + 5)^4$

then we write, $u = 2x + 5$

and express y as:

$y = u^4$

Therefore, $\dfrac{dy}{dx} = \dfrac{dy}{du} \times \dfrac{du}{dx}$

This rule is called the chain rule.

Examples

1. Find the derivatives of the following:

(a) $y = 2x^7$

(b) $y = \dfrac{3}{4}x^8$

(c) $y = 3\sqrt{x}$

(d) $y = \dfrac{10}{\sqrt[5]{x^3}}$

(e) $y = \dfrac{1}{x^{\frac{1}{4}}}$

Solution

(a) $y = 2x^7$

Solutions

(a) $y = 2x^7$

$\dfrac{dy}{dx} = 7 \times 2x^{7-1}$

This is done by multiplying the exponent (power) by the term and subtracting 1 from the exponent (power). Hence, the answer is:

$\dfrac{dy}{dx} = 14x^6$

(b) $y = \dfrac{3}{4}x^8$

Multiply the term by the exponent (power) (i.e. 8) and subtract 1 from the exponent. This gives:

$\dfrac{dy}{dx} = 8 \times \dfrac{3}{4}x^{8-1}$

$$= \frac{24}{4}x^7$$

$$\frac{dy}{dx} = 6x^7$$

(c) $y = 3\sqrt{x}$

Since $\sqrt{x} = x^{\frac{1}{2}}$, we can rewrite the expression as:

$$y = 3x^{\frac{1}{2}}$$

$$\frac{dy}{dx} = \frac{1}{2} \times 3x^{\frac{1}{2}-1}$$

$$= \frac{3}{2}x^{-\frac{1}{2}}$$

$$= \frac{3}{2} \times \frac{1}{x^{\frac{1}{2}}} \qquad \text{(Note that } x^{-\frac{1}{2}} = \frac{1}{x^{\frac{1}{2}}} \text{ from indices)}$$

$$\frac{dy}{dx} = \frac{3}{2x^{\frac{1}{2}}}$$

$$\frac{dy}{dx} = \frac{3}{2\sqrt{x}} \qquad \text{(Since } x^{\frac{1}{2}} = \sqrt{x} \text{)}$$

(d) $y = \dfrac{10}{\sqrt[5]{x^3}}$

Expressing the root in fractional form gives:

$$y = \frac{10}{(x^3)^{\frac{1}{5}}}$$

$$y = \frac{10}{x^{\frac{3}{5}}} \qquad \text{(The two exponents (powers) 3 and } \frac{1}{5} \text{ have been multiplied)}$$

Taking the denominator to the numerator changes the sign of its exponent as follows:

$$y = 10x^{-\frac{3}{5}}$$

Hence, $\dfrac{dy}{dx} = -\dfrac{3}{5} \times 10x^{-\frac{3}{5}-1}$

$$= -\frac{30}{5}x^{-\frac{8}{5}}$$

$$= -6x^{-\frac{8}{5}}$$

$$= -6 \times \frac{1}{x^{\frac{8}{5}}} \qquad \text{(Note that the inverse of a term changes the sign of its exponent)}$$

$$\frac{dy}{dx} = \frac{-6}{x^{\frac{8}{5}}}$$

Or, $\dfrac{dy}{dx} = \dfrac{-6}{\sqrt[5]{x^8}}$

(e) $y = \dfrac{1}{x^{\frac{1}{4}}}$

This can be expressed as:

$$y = x^{-\frac{1}{4}}$$

$$\dfrac{dy}{dx} = -\dfrac{1}{4} x^{-\frac{1}{4} - 1}$$

$$= -\dfrac{1}{4} x^{-\frac{5}{4}}$$

$$= -\dfrac{1}{4} \times \dfrac{1}{x^{\frac{5}{4}}}$$

$$\dfrac{dy}{dx} = -\dfrac{1}{4x^{\frac{5}{4}}}$$ (Note that the inverse of a term changes the sign of its exponent)

Or, $\dfrac{dy}{dx} = \dfrac{1}{4\sqrt[4]{x^5}}$

2. Find the derivative of each of the following:

(a) $5x^3 - 7x^2 - 3x + 8$

(b) $\dfrac{3}{5}x^5 + 2x^3 - x$

(c) $\dfrac{2x^4 - 5x^3 - 4x^2 + 3}{x^2}$

(d) $\sqrt{x} + \dfrac{1}{\sqrt{x}}$

Solutions

(a) Let the expression be $y = 5x^3 - 7x^2 - 3x + 8$

Hence, $\dfrac{dy}{dx} = \dfrac{d(5x^3)}{dx} - \dfrac{d(7x^2)}{dx} - \dfrac{d(3x)}{dx} + \dfrac{d(8)}{dx}$

This means that each part should be differentiated separately.

Hence, $\dfrac{dy}{dx} = (3 \times 5x^{3-1}) - (2 \times 7x^{2-1}) - (1 \times 3x^{1-1}) + 0$

$\dfrac{dy}{dx} = 15x^2 - 14x - 3$ (Note that $x^0 = 1$)

Note that the derivative of a constant is zero as shown by the derivative of 8

(b) $y = \frac{3}{5}x^5 + 2x^3 - x$

$\quad \frac{dy}{dx} = (5 \times \frac{3}{5}x^{5-1}) + (3 \times 2x^{3-1}) - (1 \times x^{1-1})$

$\quad\quad = 3x^4 + 6x^2 - 1$

(c) $y = \frac{2x^4 - 5x^3 - 4x^2 + 3}{x^2}$

Dividing each term in the numerator by the denominator in order to separate the expression into it different fractions gives:

$\quad y = \frac{2x^4}{x^2} - \frac{5x^3}{x^2} - \frac{4x^2}{x^2} + \frac{3}{x^2}$

$\quad\quad = 2x^2 - 5x - 4 + \frac{3}{x^2}$

$\quad y = 2x^2 - 5x - 4 + 3x^{-2}$

$\quad \frac{dy}{dx} = (2 \times 2x) - 5 - 0 + (-2 \times 3x^{-2-1})$

$\quad\quad = 4x - 5 - 6x^{-3}$

$\quad \frac{dy}{dx} = 4x - 5 - \frac{6}{x^2}$

(d) $y = \sqrt{x} + \frac{1}{\sqrt{x}}$

$\quad = x^{\frac{1}{2}} + \frac{1}{x^{\frac{1}{2}}}$

$\quad y = x^{\frac{1}{2}} + x^{-\frac{1}{2}}$

$\quad \frac{dy}{dx} = \frac{1}{2}x^{-\frac{1}{2}} - \frac{1}{2}x^{-\frac{3}{2}}$ \quad (Note that $-\frac{1}{2} - 1 = -\frac{3}{2}$)

$\quad\quad = \frac{1}{2} \times \frac{1}{x^{\frac{1}{2}}} - \frac{1}{2} \times \frac{1}{x^{\frac{3}{2}}}$

$\quad\quad = \frac{1}{2x^{\frac{1}{2}}} - \frac{1}{2x^{\frac{3}{2}}}$

$\quad \frac{dy}{dx} = \frac{1}{2\sqrt{x}} - \frac{1}{2\sqrt{x^3}}$

3. If $y = (5x - 2)^3$, find $\frac{dy}{dx}$

Solution

$\quad y = (5x - 2)^3$ \quad (This is a composite function)

188

Let us take u = 5x – 2

If 5x – 2 is replaced with U, then the question (i.e. y = (5x – 2)^3) becomes:

$$y = u^3$$

Hence, $\dfrac{dy}{du} = 3u^2$

Since u = 5x – 2

then $\dfrac{du}{dx} = 5$

Therefore, $\dfrac{dy}{dx} = \dfrac{dy}{du} \times \dfrac{du}{dx}$ (Chain rule)

$$= 3u^2 \times 5$$

$$= 15u^2$$

Now substitute 5x – 2 for u to obtain $\dfrac{dy}{dx}$ as follows:

$$\frac{dy}{dx} = 15(5x - 2)^2$$

4. If $y = \dfrac{1}{(5x^2 - 1)^4}$ find $\dfrac{dy}{dx}$

Solution

$$y = \frac{1}{(5x^2 - 1)^4}$$ (This is a composite function)

It can also be represented as follows:

$$y = (5x^2 - 1)^{-4}$$ (Its inverse changes the sign of its exponent)

Now, let us take u = 5x^2 – 1

Hence, y = u^{-4} (When 5x^2 – 1 is replaced with u in the original question)

Therefore, $\dfrac{dy}{du} = -4u^{-5}$

Since u = 5x^2 – 1

then, $\dfrac{du}{dx} = 10x$

Therefore, $\dfrac{dy}{dx} = \dfrac{dy}{du} \times \dfrac{du}{dx}$ (Chain rule)

$$= -4u^{-5} \times 10x$$

$$\frac{dy}{dx} = -40xu^{-5}$$

Now substitute 5x^2 – 1 for u. This gives:

$$\frac{dy}{dx} = -40x(5x^2 - 1)^{-5}$$

Or, $\dfrac{dy}{dx} = \dfrac{-40x}{(5x^2 - 1)^5}$ (Note the change in the sign of the exponent as it becomes

denominator)

5. If $y = (2x^3 + 7x)^{\frac{1}{2}}$ find $\dfrac{dy}{dx}$

<u>Solution</u>

$$y = (2x^3 + 7x)^{\frac{1}{2}}$$

Let $u = 2x^3 + 7x$

Hence, $y = u^{\frac{1}{2}}$

$$\frac{dy}{du} = \frac{1}{2}u^{-\frac{1}{2}}$$

Also, $u = 2x^3 + 7x$

$$\frac{du}{dx} = 6x^2 + 7$$

Therefore, $\dfrac{dy}{dx} = \dfrac{dy}{du} \times \dfrac{du}{dx}$

$$= \frac{1}{2}u^{-\frac{1}{2}} \times 6x^2 + 7$$

$$= \frac{6x^2 + 7}{2}u^{-\frac{1}{2}}$$

$$= \frac{6x^2 + 7}{2} \times \frac{1}{u^{1/2}}$$

$$\frac{dy}{dx} = \frac{6x^2 + 7}{2u^{1/2}}$$

Now, replace u with $2x^3 + 7x$. This gives:

$$\frac{dy}{dx} = \frac{6x^2 + 7}{2(2x^3 + 7x)^{1/2}}$$

Or, $\dfrac{dy}{dx} = \dfrac{6x^2 + 7}{2\sqrt{2x^3 + 7x}}$ (Note that $(2x^3 + 7x)^{\frac{1}{2}} = \sqrt{2x^3 + 7x}$)

6. Find the derivative of $3x^2 - x + 9)^4$

<u>Solution</u>

$$y = (3x^2 - x + 9)^4$$

Let $u = 3x^2 - x + 9$

Hence, $y = u^4$

$$\frac{dy}{du} = 4u^3$$

$$\frac{du}{dx} = 6x - 1$$

Therefore, $\dfrac{dy}{dx} = \dfrac{dy}{du} \times \dfrac{du}{dx}$

$$= 4u^3 \times 6x - 1$$

$$= 4(6x - 1)u^3$$

$$= (24x - 4)u^3$$

$$\frac{dy}{dx} = (24x - 4)(3x^2 - x + 9)^3 \qquad \text{(When u is replaced with } 3x^2 - x + 9\text{)}$$

7. Find the derivative of $\left(x - \dfrac{5}{x}\right)^4$

Solution

$$y = \left(x - \frac{5}{x}\right)^4$$

Let $u = x - \dfrac{5}{x}$

Therefore, $y = u^4$

$$\frac{dy}{du} = 4u^3$$

$$\frac{du}{dx} = \frac{d(x)}{dx} - \frac{d(5x^{-1})}{dx} \qquad \text{(Note that } = \frac{5}{x} = 5x^{-1}\text{)}$$

$$= 1 - (-1)5x^{-2}$$

$$= 1 + 5x^{-2}$$

$$\frac{du}{dx} = 1 + \frac{5}{x^2}$$

Therefore, $\dfrac{dy}{dx} = \dfrac{dy}{du} \times \dfrac{du}{dx}$

$$= 4u^3 \times \left(1 + \frac{5}{x^2}\right)$$

$$= \left(4 + \frac{20}{x^2}\right)u^3$$

$$\frac{dy}{dx} = \left(4 + \frac{20}{x^2}\right)\left(x - \frac{5}{x}\right)^3 \qquad \text{(When u is replaced with } \left(x - \frac{5}{x}\right)\text{)}$$

8. Differentiate with respect to x: $\dfrac{1}{2x^5 - 3x + 1}$

Solution

Let $y = \dfrac{1}{2x^5 - 3x + 1}$

Or, $y = (2x^5 - 3x + 1)^{-1} \qquad \text{(Recall from indices that } \dfrac{1}{a} = a^{-1}\text{)}$

Let us take $u = 2x^5 - 3x + 1$

Therefore, $y = u^{-1}$

$$\frac{dy}{du} = -1u^{-2}$$

$$\frac{du}{dx} = 10x^4 - 3$$

Therefore, $\dfrac{dy}{dx} = \dfrac{dy}{du} \times \dfrac{du}{dx}$

$$= -1u^{-2} \times 10x^4 - 3$$

$$= -1(10x^4 - 3)u^{-2}$$

$$= (-10x^4 + 3)u^{-2}$$

$$= \frac{3 - 10x^4}{u^2}$$

$$\frac{dy}{dx} = \frac{3 - 10x^4}{(2x^5 - 3x + 1)^2}$$

9. Differentiate with respect to x: $\sqrt{7 - 5x^3}$

Solution

Let $y = \sqrt{7 - 5x^3}$

Or, $y = (7 - 5x^3)^{\frac{1}{2}}$ (Recall from indices that $\sqrt{a} = a^{\frac{1}{2}}$)

Let $u = 7 - 5x^3$

Hence, $y = u^{\frac{1}{2}}$

$$\frac{dy}{du} = \frac{1}{2}u^{-\frac{1}{2}}$$

$$\frac{du}{dx} = -15x^2$$

Therefore, $\dfrac{dy}{dx} = \dfrac{dy}{du} \times \dfrac{du}{dx}$

$$= \frac{1}{2}u^{-\frac{1}{2}} \times (-15x^2)$$

$$= -\frac{15}{2}x^2 (u^{-\frac{1}{2}})$$

$$= -\frac{15x^2}{2u^{\frac{1}{2}}}$$

$$= -\frac{15x^2}{2\sqrt{u}}$$

$$\frac{dy}{dx} = -\frac{15x^2}{2\sqrt{7 - 5x^3}}$$

10. Find $\dfrac{dy}{dx}$ if $y = \dfrac{1}{\sqrt{2x^3 - 5}}$

Solution

$$y = \frac{1}{\sqrt{2x^3 - 5}}$$

Let $u = 2x^3 - 5$

hence, $y = \dfrac{1}{\sqrt{u}}$

$$y = \frac{1}{u^{\frac{1}{2}}}$$

$$y = u^{-\frac{1}{2}}$$

$$\frac{dy}{du} = -\frac{1}{2}u^{-\frac{3}{2}}$$ (Note that $-\frac{1}{2} - 1 = -\frac{3}{2}$)

$$\frac{du}{dx} = 6x^2$$

Therefore, $\dfrac{dy}{dx} = \dfrac{dy}{du} \times \dfrac{du}{dx}$

$$= -\frac{1}{2}u^{-\frac{3}{2}} \times (6x^2)$$

$$= -\frac{6}{2}x^2 (u^{-\frac{3}{2}})$$

$$= -3x^2(u^{-\frac{3}{2}})$$

$$= -\frac{3x^2}{u^{\frac{3}{2}}}$$

$$= -\frac{3x^2}{\sqrt{u^3}}$$

$$\frac{dy}{dx} = -\frac{3x^2}{\sqrt{(2x^3 - 5)^3}}$$

Exercise 14

1. Find the derivatives of the following:

(a) $y = 8x^5$

(b) $y = \dfrac{2}{5}x^5$

(c) $y = \sqrt[3]{x}$

(d) $y = 7\sqrt[7]{x}$

(e) $y = \dfrac{1}{\sqrt[8]{x^5}}$

(f) $y = \dfrac{2}{x^{\frac{5}{2}}}$

2. Find the derivative of each of the following:

(a) $2x^5 - 3x^4 - 4x^3 + 5x^2 - 6x + 7$

(b) $x^7 + 2x^4 - \dfrac{3}{x}$

(c) $\dfrac{3x^9 - x^7 - 5x^4 + 2x^2 - 1}{x^3}$

(d) $5(\sqrt[4]{x}) + \dfrac{5}{\sqrt[3]{2x}}$

3. If $y = (2x - 5)^4$, find $\dfrac{dy}{dx}$

4. If $y = \dfrac{3}{(x^3 - 7)^2}$ find $\dfrac{dy}{dx}$

5. If $y = (2x^3 + 7x)^{\frac{1}{2}}$ find $\dfrac{dy}{dx}$

6. Find the derivative of $(7x^3 - x^2 + 3)^5$

7. Find the derivative of $\left(3x - \dfrac{2}{3x}\right)^3$

8. Differentiate with respect to x: $\quad -\dfrac{9}{3x^2 - x - 10}$

9. Differentiate with respect to x: $\quad \sqrt{1 - 2x^4}$

10. Find $\dfrac{dy}{dx}$ if $y = \dfrac{1}{\sqrt[3]{5x^3 - 1}}$

11. Find the derivative of $(x^5 - 3)^9$

12. Find the derivative of $\left(x - \dfrac{1}{5x}\right)^2$

13. Differentiate with respect to x: $\quad \dfrac{2}{x^3 - x - \frac{1}{x}}$

14. Differentiate with respect to x: $\quad \sqrt{5x - x^2}$

15. Find $\dfrac{dy}{dx}$ if $y = \dfrac{1}{\sqrt[5]{x^3 + 2}}$

CHAPTER 15
PRODUCT RULE OF DERIVATIVE

If $y = uv$ where u and v are functions of x, then:

$$\frac{dy}{dx} = u\frac{dv}{dx} + v\frac{du}{dx}$$

This is called the product rule of differentiation.

Similarly, If $y = uvw$ where u, v and w are functions of x, then:

$$\frac{dy}{dx} = \frac{du}{dx}vw + \frac{dv}{dx}uw + \frac{dw}{dx}uv$$

Examples

1. If $f(x) = (x-3)(x+4)$, find $f'(x)$

Solution

$$f(x) = (x-3)(x+4)$$

This is a product of two functions of x.

Let $u = x - 3$

and $v = x + 4$

Hence, $\dfrac{du}{dx} = 1$

$\dfrac{dv}{dx} = 1$

Therefore, the derivative of $f(x)$ is:

$$f'(x) = u\frac{dv}{dx} + v\frac{du}{dx}$$
$$= (x-3) \times 1 + (x+4) \times 1$$
$$= x - 3 + x + 4$$
$$f'(x) = 2x + 1$$

Note that another way to differentiate the function in example 1 above is to expand the bracket and differentiate directly.

2. Find the derivative of $y = (4x^2 + 1)(x^2 - 3)$

Solution

$$y = (4x^2 + 1)(x^2 - 3)$$

Let $u = 4x^2 + 1$

and $v = (x^2 - 3)$

Hence, $\dfrac{du}{dx} = 8x$

$\dfrac{dv}{dx} = 2x$

$$\frac{dy}{dx} = u\frac{dv}{dx} + v\frac{du}{dx}$$

$$= (4x^2 + 1)2x + (x^2 - 3)8x$$

$$= 8x^3 + 2x + 8x^3 - 24x$$

$$\frac{dy}{dx} = 16x^3 - 22x$$

3. If $y = x^2(1 + 2x)^{\frac{1}{2}}$, find $\frac{dy}{dx}$

Solution

$$y = x^2(1 + 2x)^{\frac{1}{2}}$$
$$u = x^2$$

and $v = (1 + 2x)^{\frac{1}{2}}$

Hence, $\frac{du}{dx} = 2x$

$$\frac{dv}{dx} = \frac{1}{2} \times 2 \times (1 + 2x)^{\frac{1}{2}-1} \qquad \text{(Use of chain rule)}$$

$$= (1 + 2x)^{-\frac{1}{2}}$$

$$\frac{dv}{dx} = \frac{1}{(1+2x)^{\frac{1}{2}}}$$

Hence, $\frac{dy}{dx} = u\frac{dv}{dx} + v\frac{du}{dx}$

$$= x^2\frac{1}{(1+2x)^{\frac{1}{2}}} + (1 + 2x)^{\frac{1}{2}} \times 2x$$

$$= \frac{x^2}{(1+2x)^{\frac{1}{2}}} + 2x(1 + 2x)^{\frac{1}{2}}$$

let us now simplify by using $(1 + 2x)^{\frac{1}{2}}$ as the LCM as follows:

$$= \frac{x^2 + 2x\,(1+2x)^{\frac{1}{2}}(1+2x)^{\frac{1}{2}}}{(1+2x)^{\frac{1}{2}}}$$

$$= \frac{x^2 + 2x\,(1+2x)}{(1+2x)^{\frac{1}{2}}} \qquad \text{(Note that } (1 + 2x)^{\frac{1}{2}} \times (1 + 2x)^{\frac{1}{2}} = (1 + 2x)^{\frac{1}{2}+\frac{1}{2}} = 1 + 2x\text{)}$$

$$= \frac{x^2 + 2x + 4x^2}{(1+2x)^{\frac{1}{2}}}$$

$$\frac{dy}{dx} = \frac{5x^2 + 2x}{\sqrt{1+2x}}$$

4. Find the derivative of $(2x + 3)^3(4x^2 - 1)^2$

Solution

$$y = (2x + 3)^3(4x^2 - 1)^2$$

Let $u = (2x + 3)^3$

and $v = (4x^2 - 1)^2$

$\dfrac{du}{dx} = 3(2x + 3)^{3-1} \times 2$ (Use of chain rule. Also note that 2 is from the derivative of $2x + 3$)

$\qquad = 6(2x + 3)^2$

$\dfrac{dv}{dx} = 2(4x^2 - 1)^{2-1} \times 8x$ (Note that $8x$ is from the derivative of $4x^2 - 1$)

$\qquad = 16x(4x^2 - 1)$

Hence, $\dfrac{dy}{dx} = u\dfrac{dv}{dx} + v\dfrac{du}{dx}$

$\qquad = (2x + 3)^3 \times 16x(4x^2 - 1) + (4x^2 - 1)^2 \times 6(2x + 3)^2$

Let us factorize the expression by taking out $(2x + 3)^2$ and $(4x^2 - 1)$ which are the common terms as follows:

$\dfrac{dy}{dx} = (2x + 3)^2(4x^2 - 1)[(2x + 3)16x + (4x^2 - 1)6]$

$\qquad = (2x + 3)^2(4x^2 - 1)(32x^2 + 48x + 24x^2 - 6)$

$\qquad = (2x + 3)^2(4x^2 - 1)(56x^2 + 48x - 6)$

$\qquad = (2x + 3)^2(4x^2 - 1)\,2(28x^2 + 24x - 3)$

$\dfrac{dy}{dx} = 2(2x + 3)^2(4x^2 - 1)(28x^2 + 24x - 3)$

5. Differentiate: $y = x(x + 1)(x^2 - 4)$

Solution

$\qquad y = x(x + 1)(x^2 - 4)$

Expanding the first two brackets gives:

$\qquad y = (x^2 + x)(x^2 - 4)$

Hence, $u = (x^2 + x)$

and $v = (x^2 - 4)$

Hence, $\dfrac{du}{dx} = 2x + 1$

$\qquad \dfrac{dv}{dx} = 2x$

Therefore, $\dfrac{dy}{dx} = u\dfrac{dv}{dx} + v\dfrac{du}{dx}$

$\qquad = (x^2 + x)2x + (x^2 - 4)(2x + 1)$

$\qquad = 2x^3 + 2x^2 + 2x^3 + x^2 - 8x - 4$

$\dfrac{dy}{dx} = 4x^3 + 3x^2 - 8x - 4$

6. Find the derivative of $(x^2 + 3x - 2)^2 \sqrt{x}$

Solution

$\qquad y = (x^2 + 3x - 2)^2 \sqrt{x}$

$$u = (x^2 + 3x - 2)^2$$

$$v = \sqrt{x}$$

Or, $v = x^{\frac{1}{2}}$

$$\frac{du}{dx} = 2(x^2 + 3x - 2)^{2-1} \times 2x + 3 \quad \text{(Note that } 2x + 3 \text{ is from the derivative of } x^2 + 3x - 2\text{)}$$

$$= (4x + 6)(x^2 + 3x - 2)$$

$$\frac{dv}{dx} = \frac{1}{2} x^{-\frac{1}{2}} \quad \text{(Note that } 8x \text{ is from the derivative of } 4x^2 - 1\text{)}$$

$$= \frac{1}{2x^{\frac{1}{2}}}$$

$$\frac{dv}{dx} = \frac{1}{2\sqrt{x}}$$

Hence, $\dfrac{dy}{dx} = u\dfrac{dv}{dx} + v\dfrac{du}{dx}$

$$= (x^2 + 3x - 2)^2 \frac{1}{2\sqrt{x}} + \sqrt{x}\,(4x + 6)(x^2 + 3x - 2)$$

$$= (x^2 + 3x - 2)^2 \frac{1}{2\sqrt{x}} + 2\sqrt{x}\,(2x + 3)(x^2 + 3x - 2)$$

Factorize the expression by taking out $(x^2 + 3x - 2)$ which is the common factor. This gives:

$$\frac{dy}{dx} = (x^2 + 3x - 2)\left[\frac{x^2 + 3x - 2}{2\sqrt{x}} + 2\sqrt{x}\,(2x + 3) \right]$$

Simplifying the part in the bracket by taking $2\sqrt{x}$ as the LCM gives:

$$\frac{dy}{dx} = (x^2 + 3x - 2)\left[\frac{x^2 + 3x - 2 + 4x(2x + 3)}{2\sqrt{x}} \right] \quad \text{(Note that } 2\sqrt{x} \times 2\sqrt{x} = 4x\text{)}$$

$$= (x^2 + 3x - 2)\left[\frac{x^2 + 3x - 2 + 8x^2 + 12x}{2\sqrt{x}} \right]$$

$$\frac{dy}{dx} = \frac{(x^2 + 3x - 2)(9x^2 + 15x - 2)}{2\sqrt{x}}$$

7. Find the derivative of $(1 + x)(5x - 2)^{\frac{3}{2}}$

<u>Solution</u>

$$y = (1 + x)(5x - 2)^{\frac{3}{2}}$$

$$u = (1 + x)$$

$$v = (5x - 2)^{\frac{3}{2}}$$

$$\frac{du}{dx} = 1$$

$$\frac{dv}{dx} = \frac{3}{2}(5x - 2)^{\frac{3}{2} - 1} \times 5 \quad \text{(Note that the derivative of } 5x - 2 \text{ is 5)}$$

$$= \frac{15}{2}(5x - 2)^{\frac{1}{2}}$$

Hence, $\dfrac{dy}{dx} = u\dfrac{dv}{dx} + v\dfrac{du}{dx}$

$$= (1 + x) \times \frac{15}{2} (5x - 2)^{\frac{1}{2}} + (5x - 2)^{\frac{3}{2}} \times 1$$

$$= \frac{15}{2} (1 + x)(5x - 2)^{\frac{1}{2}} + (5x - 2)^{\frac{3}{2}}$$

Factorize by taking out $(5x - 2)^{\frac{1}{2}}$ (i.e. the lower exponent) which is the common factor gives:

$$\frac{dy}{dx} = (5x - 2)^{\frac{1}{2}} \left[\frac{15}{2} (1 + x) + 5x - 2 \right] \quad \text{(Note that } \frac{(5x-2)^{\frac{3}{2}}}{(5x-2)^{\frac{1}{2}}} = (5x - 2)^{\frac{3}{2} - \frac{1}{2}} = 5x - 2\text{)}$$

$$= (5x - 2)^{\frac{1}{2}} \left[\frac{15}{2} + \frac{15x}{2} + 5x - 2 \right]$$

$$= (5x - 2)^{\frac{1}{2}} \left[\frac{25x}{2} + \frac{11}{2} \right]$$

$$\frac{dy}{dx} = \sqrt{5x - 2} \left[\frac{25x + 11}{2} \right]$$

8. If y = (1 + x)(2 − 3x)(2x − 1), find $\dfrac{dy}{dx}$ by using product rule.

Solution

$\quad$ y = (1 + x)(2 − 3x)(2x − 1)

This is a product of three expressions, u, v and w.

Hence, u = (1 + x)

$\qquad$ v = (2 − 3x)

and $\quad$ w = (2x − 1)

Therefore, $\dfrac{du}{dx} = 1$

$\dfrac{dv}{dx} = -3$

$\dfrac{dw}{dx} = 2$

Hence the formula for product rule of three terms is given by:

$$\frac{dy}{dx} = \frac{du}{dx} vw + \frac{dv}{dx} uw + \frac{dw}{dx} uv$$

$$= 1(2 - 3x)(2x - 1) + (-3)(1 + x)(2x - 1) + 2(1 + x)(2 - 3x)$$

$$= 4x - 2 - 6x^2 + 3x + (-3 - 3x)(2x - 1) + (2 + 2x)(2 - 3x)$$

$$= 7x - 2 - 6x^2 - 6x + 3 - 6x^2 + 3x + 4 - 6x + 4x - 6x^2$$

$$= -2 + 3 + 4 + 7x - 6x + 3x - 6x + 4x - 6x^2 - 6x^2 - 6x^2$$

$$\frac{dy}{dx} = 5 + 2x - 18x^2$$

9. Differentiate with respect to x: $(x^2 - 3x + 5)(2x - 7)$

Solution

$\quad$ y = $(x^2 - 3x + 5)(2x - 7)$

Let us differentiate this product by applying product rule but without the use u and v. This is

199

done as follows:

$$\frac{dy}{dx} = (x^2 - 3x + 5)\frac{d(2x-7)}{dx} + (2x - 7)\frac{d(x^2 - 3x + 5)}{dx}$$

$$= (x^2 - 3x + 5)2 + (2x - 7)(2x - 3)$$

$$= 2x^2 - 6x + 10 + 4x^2 - 6x - 14x + 21$$

$$\frac{dy}{dx} = 6x^2 - 26x + 31$$

10. If $y = (5x^2 - 3)(2 + \frac{3}{x})$, find $\frac{dy}{dx}$

Solution

$$y = (5x^2 - 3)(2 + \frac{3}{x})$$

$$u = (5x^2 - 3)$$

$$v = (2 + \frac{3}{x})$$

$$\frac{du}{dx} = 10x$$

$$\frac{dv}{dx} = \frac{d(3x^{-1})}{dx} \qquad \text{(Note that } \frac{3}{x} = 3x^{-1}\text{)}$$

$$= -3x^{-2}$$

$$\frac{dv}{dx} = \frac{-3}{x^2}$$

Hence, $\quad \frac{dy}{dx} = u\frac{dv}{dx} + v\frac{du}{dx}$

$$= (5x^2 - 3)\left(\frac{-3}{x^2}\right) + (2 + \frac{3}{x})10x$$

$$= -15 + \frac{9}{x^2} + 20x + 30$$

$$\frac{dy}{dx} = 15 + 20x + \frac{9}{x^2}$$

Exercise 15

1. If $f(x) = (2x - 1)(3x + 1)$, find $f'(x)$

2. Find the derivative of $y = (3x^2 - 5)(x^2 + 10)$

3. If $y = 5x(3 + x)^{\frac{1}{2}}$ find $\frac{dy}{dx}$

4. Find the derivative of $(x + 5)(x^2 - 7)^3$

5. Differentiate: $y = 2x(3x + 2)(2x^2 - 5)$

6. Find the derivative of $(3x^2 - 1)^3 \sqrt{2x}$

7. Find the derivative of $(9 - x)(x + 3)^{\frac{3}{4}}$

8. If $y = (2 - x)(5 - 3x^2)(x + 3)$, find $\dfrac{dy}{dx}$ by using product rule.

9. Differentiate with respect to x: $(3x^4 - x^2 + 2x)(x - 1)$

10. If $y = (3x^2 - x)(1 + \dfrac{1}{2x})$, find $\dfrac{dy}{dx}$

11. Find the derivative of $x^2(\sqrt{x^5})$

12. Find the derivative of $x^4(2x - 11)^{\frac{2}{3}}$

13. If $y = (7 + x)(1 - x^2)(5x^3 + 1)$, find $\dfrac{dy}{dx}$.

14. Differentiate with respect to x: $(3x^4 - x^3 + x^2 + 2x - 3)(5x + 4)$

15. If $y = (x^3 - 3x + 5)\left(\dfrac{1}{x^5}\right)$, find $\dfrac{dy}{dx}$

CHAPTER 16
QUOTIENT RULE OF DERIVATIVE

If $y = \dfrac{u}{v}$ where u and v are functions of x, then:

$$\frac{dy}{dx} = \frac{v\frac{du}{dx} - u\frac{dv}{dx}}{v^2}$$

This is called the quotient rule of differentiation.

Examples

1. If $y = \dfrac{3x^2 - 8x + 5}{5x - 2}$ find $\dfrac{dy}{dx}$

Solution

$$y = \frac{3x^2 - 8x + 5}{5x - 2}$$

This is of the form $y = \dfrac{u}{v}$. Therefore, we are going to apply product rule.

Let $u = 3x^2 - 8x + 5$

and $v = 5x - 2$

Hence, $\dfrac{du}{dx} = 6x - 8$

$\dfrac{dv}{dx} = 5$

Therefore, $\dfrac{dy}{dx} = \dfrac{v\frac{du}{dx} - u\frac{dv}{dx}}{v^2}$

$$= \frac{(5x - 2)(6x - 8) - (3x^2 - 8x + 5)(5)}{(5x - 2)^2}$$

$$= \frac{30x^2 - 40x - 12x + 16 - 15x^2 - 40x - 25}{(5x - 2)^2}$$

$$\frac{dy}{dx} = \frac{15x^2 - 12x - 9}{(5x - 2)^2}$$

2. Differentiate with respect to x, the function: $\dfrac{3x^2 - 2x}{x + 5}$

Solution

$$y = \frac{3x^2 - 2x}{x + 5}$$

Let $u = 3x^2 - 2x$

and $v = x + 5$

Hence, $\dfrac{du}{dx} = 6x - 2$

$\dfrac{dv}{dx} = 1$

Therefore, $\dfrac{dy}{dx} = \dfrac{v\frac{du}{dx} - u\frac{dv}{dx}}{v^2}$

$= \dfrac{(x+5)(6x-2) - (3x^2 - 2x)(1)}{(x+5)^2}$

$= \dfrac{6x^2 - 2x + 30x - 10 - 3x^2 + 2x}{(x+5)^2}$

$\dfrac{dy}{dx} = \dfrac{3x^2 + 30x - 10}{(x+5)^2}$

3. Find the differential coefficient of $y = \dfrac{-3}{x^2 + 5}$

<u>Solution</u>

$y = \dfrac{-3}{x^2 + 5}$

$u = -3$

and $v = x^2 + 5$

Hence, $\dfrac{du}{dx} = 0$ (The derivative of a constant is zero)

$\dfrac{dv}{dx} = 2x$

Therefore, $\dfrac{dy}{dx} = \dfrac{v\frac{du}{dx} - u\frac{dv}{dx}}{v^2}$

$= \dfrac{(x^2 + 5) \times 0 - (-3)2x}{(x^2 + 5)^2}$

$= \dfrac{0 - 6x}{(x^2 + 5)^2}$

$\dfrac{dy}{dx} = \dfrac{-6x}{(x^2 + 5)^2}$

4. Differentiate: $y = \dfrac{\sqrt{3-x}}{\sqrt{3+x}}$

<u>Solution</u>

$y = \dfrac{\sqrt{3-x}}{\sqrt{3+x}}$

$u = \sqrt{3-x}$

$= (3-x)^{\frac{1}{2}}$

and $v = \sqrt{3+x}$

$= (3+x)^{\frac{1}{2}}$

Hence, $\dfrac{du}{dx} = \dfrac{1}{2}(3-x)^{\frac{1}{2}-1} \times -1$ (Note that the derivative of $3-x$ is -1)

$= -\dfrac{1}{2}(3-x)^{-\frac{1}{2}}$

203

$$\frac{du}{dx} = \frac{-(3-x)^{-\frac{1}{2}}}{2}$$

$$\frac{dv}{dx} = \frac{1}{2}(3+x)^{\frac{1}{2}-1} \times 1 \quad \text{(Note that the derivative of } 3-x \text{ is } -1)$$

$$= \frac{1}{2}(3+x)^{-\frac{1}{2}}$$

$$\frac{dv}{dx} = \frac{(3+x)^{-\frac{1}{2}}}{2}$$

Therefore, $\dfrac{dy}{dx} = \dfrac{v\frac{du}{dx} - u\frac{dv}{dx}}{v^2}$

$$= \frac{(3+x)^{\frac{1}{2}}\left(\frac{-(3-x)^{-\frac{1}{2}}}{2}\right) - (3-x)^{\frac{1}{2}}\left(\frac{(3+x)^{-\frac{1}{2}}}{2}\right)}{\left((3+x)^{\frac{1}{2}}\right)^2}$$

Taking out the common terms which are terms with positive exponent in order to factorize the expression gives:

$$\frac{dy}{dx} = \frac{\left((3+x)^{\frac{1}{2}}\right)\left((3-x)^{\frac{1}{2}}\right)\left[\frac{-(3-x)^{-1}}{2} - \frac{(3+x)^{-1}}{2}\right]}{3+x}$$

Note that in order to obtain the terms in the square bracket, we subtracted the exponents of the factors from the exponents of the original term. For example, $(3+x)^{\frac{1}{2}}\left(\frac{-(3-x)^{-\frac{1}{2}}}{2}\right)$ divided

by $\left((3+x)^{\frac{1}{2}}\right)\left((3-x)^{\frac{1}{2}}\right)$ gave $\dfrac{-(3-x)^{-1}}{2}$ since the equal terms canceled out and $-\frac{1}{2} - \frac{1}{2} = -1$,

which gave the exponent of -1. Similarly, $(3-x)^{\frac{1}{2}}\left(\frac{(3+x)^{-\frac{1}{2}}}{2}\right)$ divided by

$\left((3+x)^{\frac{1}{2}}\right)\left((3-x)^{\frac{1}{2}}\right)$ gave $\dfrac{(3+x)^{-1}}{2}$ since the equal terms canceled out and $-\frac{1}{2} - \frac{1}{2} = -1$, which

gave the exponent of -1 as the terms in the square bracket.

Let us now continue with the solution by simplifying further as follows:

$$\frac{dy}{dx} = \frac{\left((3+x)^{\frac{1}{2}}\right)\left((3-x)^{\frac{1}{2}}\right)\left[\frac{-(3-x)^{-1}}{2} - \frac{(3+x)^{-1}}{2}\right]}{3+x}$$

$$= \frac{\left((3+x)^{\frac{1}{2}}\right)\left((3-x)^{\frac{1}{2}}\right)\left[\frac{-1}{2(3-x)} - \frac{1}{2(3+x)}\right]}{3+x}$$

$$= \frac{\left((3+x)^{\frac{1}{2}}\right)\left((3-x)^{\frac{1}{2}}\right)\left[\frac{-(3+x) - (3-x)}{2(3-x)(3+x)}\right]}{3+x}$$

$$= \frac{\left((3+x)^{\frac{1}{2}}\right)\left((3-x)^{\frac{1}{2}}\right)\left[\frac{-3-x-3+x}{2(3-x)(3+x)}\right]}{3+x}$$

$$= \frac{\left((3+x)^{\frac{1}{2}}\right)\left((3-x)^{\frac{1}{2}}\right)\left[\frac{-6}{2(3-x)(3+x)}\right]}{3+x}$$

$$= \frac{\left((3+x)^{\frac{1}{2}}\right)\left((3-x)^{\frac{1}{2}}\right)\left[\frac{-3}{(3-x)(3+x)}\right]}{3+x}$$

$$= \frac{-3\left((3+x)^{\frac{1}{2}}\right)\left((3-x)^{\frac{1}{2}}\right)}{(3-x)(3+x)(3+x)}$$

$$= \frac{-3\left((3+x)^{\frac{1}{2}}\right)\left((3-x)^{\frac{1}{2}}\right)}{(3+x)^2(3-x)}$$

$$= -3(3+x)^{\frac{1}{2}-2}(3-x)^{\frac{1}{2}-1} \qquad \text{(Subtraction of exponents due to the division above)}$$

$$= -3(3+x)^{-\frac{3}{2}}(3-x)^{-\frac{1}{2}}$$

$$= \frac{-3}{(3+x)^{\frac{3}{2}}(3-x)^{\frac{1}{2}}}$$

$$= \frac{-3}{\sqrt{(3+x)^3(3-x)}}$$

5. Differentiate with respect to x: $y = \dfrac{(2x^2-3)^3}{x}$

<u>Solution</u>

$$y = \frac{(2x^2-3)^3}{x}$$
$$u = (2x^2-3)^3$$

and $v = x$

Hence, $\dfrac{du}{dx} = 3(2x^2-3)^{3-1} \times 4x$

$$= 12x(2x^2-3)^2$$

$$\frac{dv}{dx} = 1$$

Therefore, $\dfrac{dy}{dx} = \dfrac{v\frac{du}{dx} - u\frac{dv}{dx}}{v^2}$

$$= \frac{x[12x(2x^2-3)^2] - (2x^2-3)^3 \times 1}{x^2}$$

$$= \frac{12x^2(2x^2-3)^2 - (2x^2-3)^3}{x^2}$$

Factorize the expression by taking out $(2x^2-3)^2$ which has the lower exponent. This gives:

$$\frac{dy}{dx} = \frac{(2x^2 - 3)^2 [12x^2 - (2x^2 - 3)]}{x^2}$$

$$= \frac{(2x^2 - 3)^2 (12x^2 - 2x^2 + 3)}{x^2}$$

$$\frac{dy}{dx} = \frac{(2x^2 - 3)^2 (10x^2 + 3)}{x^2}$$

6. Find the derivative of $\dfrac{\sqrt{(1 + 2x^2)^3}}{x}$

Solution

$$y = \frac{\sqrt{(1 + 2x^2)^3}}{x}$$

$$u = \sqrt{(1 + 2x^2)^3}$$

$$= [(1 + 2x^2)^3]^{\frac{1}{2}} \qquad \text{(When the square root sign is removed, we use an exponent of } \frac{1}{2}\text{)}$$

$$u = (1 + 2x^2)^{\frac{3}{2}} \qquad \text{(After multiplying the exponents)}$$

$$v = x$$

Hence, $\dfrac{du}{dx} = \dfrac{3}{2} (1 + 2x^2)^{\frac{3}{2} - 1} \times 4x$ (Note that the derivative of $3 - x$ is -1)

$$= 6x(1 + 2x^2)^{\frac{1}{2}}$$

$$\frac{dv}{dx} = 1$$

Therefore, $\dfrac{dy}{dx} = \dfrac{v\frac{du}{dx} - u\frac{dv}{dx}}{v^2}$

$$= \frac{x\left(6x(1 + 2x^2)^{\frac{1}{2}}\right) - (1 + 2x^2)^{\frac{3}{2}} \times 1}{x^2}$$

Take out $(1 + 2x^2)^{\frac{1}{2}}$ which has the lower exponent and factorize the expression. This gives:

$$\frac{dy}{dx} = \frac{(1 + 2x^2)^{\frac{1}{2}} [6x^2 - (1 + 2x^2)]}{x^2} \qquad \text{(Note that } \frac{(1 + 2x^2)^{\frac{3}{2}}}{(1 + 2x^2)^{\frac{1}{2}}} \text{ gives } (1 + 2x^2) \text{ by subtracting exponents.}$$

$$= \frac{(1 + 2x^2)^{\frac{1}{2}} (6x^2 - 1 - 2x^2)}{x^2}$$

$$= \frac{(1 + 2x^2)^{\frac{1}{2}} (4x^2 - 1)}{x^2}$$

$$\frac{dy}{dx} = \frac{(\sqrt{1 + 2x^2})(4x^2 - 1)}{x^2}$$

7. If $y = \dfrac{(4x^3 - 3x^2 + x + 1)^{\frac{1}{2}}}{(x+1)^2}$

Solution

$$y = \dfrac{(4x^3 - 3x^2 + x + 1)^{\frac{1}{2}}}{(x+1)^2}$$

$u = (4x^3 - 3x^2 + x + 1)^{\frac{1}{2}}$ (After multiplying the exponents)

$v = (x + 1)^2$

Hence, $\dfrac{du}{dx} = \dfrac{1}{2}(4x^3 - 3x^2 + x + 1)^{\frac{1}{2} - 1} \times (12x^2 - 6x + 1)$

$$= \dfrac{(12x^2 - 6x + 1)(4x^3 - 3x^2 + x + 1)^{-\frac{1}{2}}}{2}$$

$\dfrac{dv}{dx} = 2(x + 1)^{2-1}$

$\quad = 2(x + 1)$

Therefore, $\dfrac{dy}{dx} = \dfrac{v\dfrac{du}{dx} - u\dfrac{dv}{dx}}{v^2}$

$$= \dfrac{\dfrac{(x+1)^2(12x^2 - 6x + 1)(4x^3 - 3x^2 + x + 1)^{-\frac{1}{2}} - (4x^3 - 3x^2 + x + 1)^{\frac{1}{2}}[2(x+1)]}{2}}{[(x+1)^2]^2}$$

Take out the common terms with lower exponents [i.e. $(x + 1)$ and $(4x^3 - 3x^2 + x + 1)^{-\frac{1}{2}}$] and factorize the expression. This gives:

$$\dfrac{dy}{dx} = \dfrac{\dfrac{(x+1)(4x^3 - 3x^2 + x + 1)^{-\frac{1}{2}}[(x+1)(12x^2 - 6x + 1) - (4x^3 - 3x^2 + x + 1)2]}{2}}{[(x+1)^2]^2}$$

Remember to subtract the exponents of the factors from the exponents of the original expression when simplifying. Simplifying further, the expression above gives:

$$\dfrac{dy}{dx} = \dfrac{(x+1)(4x^3 - 3x^2 + x + 1)^{-\frac{1}{2}}[12x^3 - 6x^2 + x + 12x^2 - 6x + 1 - (8x^3 - 6x^2 + 2x + 2)]}{2(x+1)^4}$$

$$= \dfrac{(x+1)(4x^3 - 3x^2 + x + 1)^{-\frac{1}{2}}(12x^3 - 6x^2 + x + 12x^2 - 6x + 1 - 8x^3 + 6x^2 - 2x - 2)}{2(x+1)^4}$$

$$= \dfrac{(x+1)(4x^3 - 3x^2 + x + 1)^{-\frac{1}{2}}(4x^3 + 12x^2 - 7x - 1)}{2(x+1)^4}$$

$$= \dfrac{(4x^3 - 3x^2 + x + 1)^{-\frac{1}{2}}(4x^3 + 12x^2 - 7x - 1)}{2(x+1)^3}$$

Note that $(x + 1)$ cancels out from the numerator and denominator.

Therefore, $\dfrac{dy}{dx} = \dfrac{4x^3 + 12x^2 - 7x - 1}{2(4x^3 - 3x^2 + x + 1)^{\frac{1}{2}}(x+1)^3}$

8. Determine $\dfrac{d}{dx}\left(\dfrac{3 + 2x - x^2}{\sqrt{1+x}}\right)$

Solution

Let $y = \dfrac{3 + 2x - x^2}{\sqrt{1+x}}$

Hence, $u = 3 + 2x - x^2$

$\qquad \dfrac{du}{dx} = 2 - 2x$

$\qquad v = \sqrt{1 + x}$

$\qquad = (1 + x)^{\frac{1}{2}}$

$\qquad \dfrac{du}{dx} = \dfrac{1}{2}(1 + x)^{\frac{1}{2} - 1} \times 1$

$\qquad = \dfrac{1}{2}(1 + x)^{-\frac{1}{2}}$

Therefore, $\dfrac{dy}{dx} = \dfrac{v\frac{du}{dx} - u\frac{dv}{dx}}{v^2}$

$= \dfrac{(1 + x)^{\frac{1}{2}}(2 - 2x) - (3 + 2x - x^2)\frac{1}{2}(1 + x)^{-\frac{1}{2}}}{[(1 + x)^{\frac{1}{2}}]^2}$

Take out $(1 + x)^{-\frac{1}{2}}$ as the common factor since it has the lower exponent and factorize the expression. This gives:

$\dfrac{dy}{dx} = \dfrac{(1 + x)^{-\frac{1}{2}}\left[(1+x)(2-2x) - (3 + 2x - x^2)\frac{1}{2}\right]}{1+x}$

$= \dfrac{(1 + x)^{-\frac{1}{2}}\left[\dfrac{2(1+x)(2-2x) - (3 + 2x - x^2)}{2}\right]}{1+x}$

$= \dfrac{(1 + x)^{-\frac{1}{2}}[(2 + 2x)(2 - 2x) - 3 - 2x + x^2]}{2(1+x)}$

$= \dfrac{(1 + x)^{-\frac{1}{2}}(4 - 4x + 4x - 4x^2 - 3 - 2x + x^2)}{2(1+x)}$

$$= \frac{1 - 2x - 3x^2}{2(1+x)^{\frac{1}{2}}(1+x)} \qquad \text{[When } (1+x)^{-\frac{1}{2}} \text{ is taken to the denominator it becomes } (1+x)^{\frac{1}{2}}]$$

$$\frac{dy}{dx} = \frac{1 - 2x - 3x^2}{2(1+x)^{\frac{3}{2}}} \qquad \text{(The exponents of same terms have been added together, i.e. } \frac{1}{2} + 1 = \frac{3}{2})$$

Exercise 16

1. If $y = \dfrac{x^2 - 5x + 1}{x - 1}$ find $\dfrac{dy}{dx}$

2. Differentiate with respect to x, the function: $\dfrac{4x^2 - x}{2x + 3}$

3. Find the differential coefficient of $y = \dfrac{-7}{3x^2 + 1}$

4. Differentiate: $y = \dfrac{\sqrt{1+x}}{\sqrt{1-x}}$

5. Differentiate with respect to x: $y = \dfrac{(x^3 - 2)^2}{x^2}$

6. Find the derivative of $\dfrac{\sqrt{2 + 3x^2}}{x^3}$

7. If $y = \dfrac{(x^2 - x - 4)^{\frac{1}{3}}}{(2x+1)^2}$

8. Determine $\dfrac{d}{dx}\left(\dfrac{x - 2x^3}{\sqrt{2-x}}\right)$

9. Find the differential coefficient of $y = -\dfrac{1}{1 - 3x^2}$

10. Differentiate: $y = \dfrac{2 + x}{2 - x}$

CHAPTER 17
DERIVATIVE OF PARAMETRIC EQUATIONS

If y = f(t) and x = g(t) are two different functions of a common variable, t, then the two equations are called parametric equations. The variable t, is the parameter.

The derivative of a parametric equation such as the one stated above is obtained as follows:

$$\frac{dy}{dx} = \frac{\frac{dy}{dt}}{\frac{dx}{dt}}$$

Examples

1. If y = 5 + t^2 and x = 3 – 2t^2, find $\frac{dy}{dx}$

Solution

$$y = 5 + t^2$$

Hence, $\frac{dy}{dt}$ = 2t

$$x = 3 - 2t^2$$

Hence, $\frac{dx}{dt}$ = – 4t

Therefore, $\frac{dy}{dx} = \frac{\frac{dy}{dt}}{\frac{dx}{dt}}$

$$= \frac{2t}{-4t}$$

$$= -\frac{1}{2} \qquad \text{(t cancels out)}$$

2. Find $\frac{dy}{dx}$ of the functions below which are expressed in the parametric form.

$$x = \frac{5}{t^2} \text{ and } y = 2t^5 - 3$$

Solution

$$x = \frac{5}{t^2}$$

$$x = 5t^{-2}$$

$$\frac{dx}{dt} = -10t^{-3}$$

$$y = 2t^5 - 3$$

$$\frac{dy}{dt} = 10t^4$$

Therefore, $\frac{dy}{dx} = \frac{\frac{dy}{dt}}{\frac{dx}{dt}}$

$$= \frac{10t^4}{-10t^{-3}}$$
$$= -t^{4-(-3)} \qquad \text{(10 cancels out)}$$
$$= -t^{4+3}$$
$$= -t^7$$

3. If $V = \frac{4}{3}\pi r^3$ and $A = \pi r^2$, find $\frac{dA}{dV}$

Solution

$$V = \frac{4}{3}\pi r^3$$
$$\frac{dV}{dr} = 3(\frac{4}{3})\pi r^2$$
$$= 4\pi r^2$$
$$A = \pi r^2$$
$$\frac{dA}{dr} = 2\pi r$$
$$\frac{dA}{dV} = \frac{\frac{dA}{dr}}{\frac{dV}{dr}}$$
$$= \frac{2\pi r}{4\pi r^2}$$
$$\frac{dA}{dV} = \frac{1}{2r}$$

4. Determine the derivative of the curve defined by the equations: $x = t^2 - 4t$ and $y = 2t^3 - 7t$.

Solution

$$y = 2t^3 - 7t$$
Hence, $\frac{dy}{dt} = 6t^2 - 7$
$$x = t^2 - 4t$$
Hence, $\frac{dx}{dt} = 2t - 4$

Therefore, $\frac{dy}{dx} = \frac{\frac{dy}{dt}}{\frac{dx}{dt}}$

$$= \frac{6t^2 - 7}{2t - 4}$$

5. Given that $v = u + at$ and $s = ut + \frac{1}{2}at^2$. Find $\frac{dv}{ds}$ if u and a are constants.

Solution

$$v = u + at$$

$$\frac{dv}{dt} = a$$

$$s = ut + \frac{1}{2}at^2$$

$$\frac{ds}{dt} = u + (2 \times \frac{1}{2} \times at)$$

$$= u + at$$

Hence, $\dfrac{dv}{ds} = \dfrac{\frac{dv}{dt}}{\frac{ds}{dt}}$

$$= \frac{a}{u+at}$$

6. If $y = \dfrac{t}{t-2}$ and $x = \dfrac{1}{t+1}$, find $\dfrac{dy}{dx}$

Solution

$$y = \frac{t}{t-2}$$

$$\frac{dy}{dt} = \frac{(t-2)(1) - t(1)}{(t-2)^2} \qquad \text{(Use of quotient rule where u = t and v = t - 2)}$$

$$= \frac{t-2-t}{(t-2)^2}$$

$$\frac{dy}{dt} = \frac{-2}{(t-2)^2}$$

$$x = \frac{1}{t+1}$$

$$\frac{dx}{dt} = \frac{(t+1)(0) - 1(1)}{(t+1)^2} \qquad \text{(Use of quotient rule where u = 1 and v = t + 1)}$$

$$= \frac{-1}{(t+1)^2}$$

Hence, $\dfrac{dy}{dx} = \dfrac{\frac{dy}{dt}}{\frac{dx}{dt}}$

$$= \frac{\frac{-2}{(t-2)^2}}{\frac{-1}{(t+1)^2}}$$

$$= \frac{-2}{(t-2)^2} \times \frac{(t+1)^2}{-1}$$

$$\frac{dy}{dx} = \frac{2(t+1)^2}{(t-2)^2}$$

7. The parametric equations of the motion of a stone are: $y = 12 + 3t - 2t^2$ and $x = 5t$. Find $\dfrac{dy}{dx}$.

Solution

$$y = 12 + 3t - 2t^2$$

$$\frac{dy}{dt} = 3 - 4t$$

$$x = 5t$$

$$\frac{dx}{dt} = 5$$

$$\frac{dy}{dx} = \frac{\frac{dy}{dt}}{\frac{dx}{dt}}$$

$$= \frac{3 - 4t}{5}$$

8. If the parametric equations of a parabola are $y = \dfrac{2mt^2}{1+t^2}$ and $x = \dfrac{2m}{1+t^2}$ where m is a constant, find $\dfrac{dy}{dx}$.

Solution

$$y = \frac{2mt^2}{1+t^2}$$

$$\frac{dy}{dt} = \frac{(1+t^2)(4mt) - 2mt^2(2t)}{(1+t^2)^2} \qquad \text{(Use of quotient rule where u = } 2mt^2 \text{ and v = } 1 + t^2)$$

$$= \frac{4mt + 4mt^3 - 4mt^3}{(1+t^2)^2}$$

$$\frac{dy}{dt} = \frac{4mt}{(1+t^2)^2}$$

$$x = \frac{2m}{1+t^2}$$

$$\frac{dx}{dt} = \frac{(1+t^2)(0) - 2m(2t)}{(1+t^2)^2} \qquad \text{(Use of quotient rule where u = 1 and v = t + 1)}$$

$$= \frac{-4mt}{(1+t^2)^2}$$

Hence, $\dfrac{dy}{dx} = \dfrac{\frac{dy}{dt}}{\frac{dx}{dt}}$

$$= \frac{\frac{4mt}{(1+t^2)^2}}{\frac{-4mt}{(1+t^2)^2}}$$

$$= \frac{4mt}{(1+t^2)^2} \times \frac{(1+t^2)^2}{-4mt}$$

$$\frac{dy}{dx} = -1 \qquad \text{(Same terms cancels out)}$$

Exercise 17

1. If $y = t^3 - 5$ and $x = t^2 + 1$ find $\dfrac{dy}{dx}$

2. Find $\dfrac{dy}{dx}$ of the functions below which are expressed in the parametric form:

   $x = \dfrac{1}{4 - t^3}$ and $y = 3 + t^2$

3. If $V = \dfrac{1}{3}\pi r^2$ and $A = \pi r l + \pi r^2$, find $\dfrac{dV}{dA}$

4. Find the derivative of the curve defined by the equations: $x = 5t^2 - t + 3$ and $y = t^2 - t - 1$.

5. Given that $F = \dfrac{m(v - u)}{t}$ and $s = \dfrac{(u+v)t}{2}$. Find $\dfrac{dF}{ds}$ if u, v and m are constants.

6. If $y = \dfrac{5}{t^2 - 1}$ and $x = \dfrac{3}{t^4 + 1}$, find $\dfrac{dy}{dx}$

7. The parametric equations of the motion of a stone are: $y = t - 3t^4$ and $x = t^2 + 3$. Find $\dfrac{dy}{dx}$.

8. If the parametric equations of a parabola are $y = \dfrac{at^3}{2}$ and $x = \dfrac{at}{5}$ where a is a constant, find $\dfrac{dy}{dx}$.

9. If $V = \dfrac{1}{3}\pi r^3$ and $A = 4\pi r^2$, find $\dfrac{dA}{dV}$

10. Given that $G = 2m + s^2$ and $H = m^2 - \dfrac{3}{s}$. Find $\dfrac{dG}{dH}$ if m is a constant.

CHAPTER 18
DERIVATIVE OF IMPLICIT FUNCTIONS

In the function y = f(x), y is said to be expressed explicitly in terms of x. However, in expressions such as $2xy - x^2y = 5$, the relationship between y and x is said to be implicit.

In order to differentiate implicit functions, y is differentiated just like x but with the addition of $\frac{dy}{dx}$ along with the value obtained.

Examples

1. Differentiate implicitly, the expression: $2x^2 + y^2 = 9$.

<u>Solution</u>

$$2x^2 + y^2 = 9$$

Follow the rule of differentiation and add $\frac{dy}{dx}$ to the value obtained whenever you differentiate y. Hence we differentiate each term in the expression above as follows:

$$\frac{d(2x^2)}{dx} + \frac{d(2y^2)}{dx} = \frac{d(9)}{dx}$$

$$4x + 2y\frac{dy}{dx} = 0 \quad \text{(The derivative of } y^2 \text{ is 2y and the addition of } \frac{dy}{dx} \text{ to 2y gives } 2y\frac{dy}{dx}\text{)}$$

We now make $\frac{dy}{dx}$ the subject of the formula as follows:

$$2y\frac{dy}{dx} = -4x$$

$$\frac{dy}{dx} = \frac{-4x}{2y} \quad \text{(When both sides are divided by 2y)}$$

$$\frac{dy}{dx} = \frac{-2x}{y}$$

2. If $x^3 + y^3 = 18xy$, find $\frac{dy}{dx}$

<u>Solution</u>

$$x^3 + y^3 = 18xy$$

$$3x^2 + 3y^2\frac{dy}{dx} = (18x \times 1\frac{dy}{dx}) + (y \times 18)$$

Note that 18xy is differentiated by using product rule where 18x is taken as u while y is take as

v. Also, the derivative of y is what gave us $1\frac{dy}{dx}$. The above differentiation now simplifies to:

$$3x^2 + 3y^2\frac{dy}{dx} = 18x\frac{dy}{dx} + 18y$$

Collect terms in $\frac{dy}{dx}$ on one side in order to make $\frac{dy}{dx}$ the subject of the formula as follows:

$$3y^2\frac{dy}{dx} - 18x\frac{dy}{dx} = 18y - 3x^2$$

Factorizing the left hand side gives:

$$\frac{dy}{dx}(3y^2 - 18x) = 18y - 3x^2$$

Divide both sides by $3y^2 - 18x$. This gives:

$$\frac{dy}{dx} = \frac{18y - 3x^2}{3y^2 - 18x}$$

$$= \frac{3(6y - x^2)}{3(y^2 - 6x)}$$

$$\frac{dy}{dx} = \frac{6y - x^2}{y^2 - 6x} \qquad \text{(3 cancels out)}$$

3. Find $\frac{dy}{dx}$ given that $x^2y^2 - 3xy + 4xy^3 = 4$

Solution

$$x^2y^2 - 3xy + 4xy^3 = 4$$

Apply product rule to x^2y^2, $3xy$ and $4xy^3$ and differentiate appropriately as follows:

$$(x^2 \times 2y\frac{dy}{dx}) + (y^2 \times 2x) - [(3x \times 1\frac{dy}{dx}) + (y \times 3)] + (4x \times 3y^2\frac{dy}{dx}) + (y^3 \times 4) = 0$$

$$2x^2y\frac{dy}{dx} + 2xy^2 - (3x\frac{dy}{dx} + 3y) + 12xy^2\frac{dy}{dx} + 4y^3 = 0$$

$$2x^2y\frac{dy}{dx} + 2xy^2 - 3x\frac{dy}{dx} - 3y + 12xy^2\frac{dy}{dx} + 4y^3 = 0$$

$$2x^2y\frac{dy}{dx} - 3x\frac{dy}{dx} + 12xy^2\frac{dy}{dx} + 2xy^2 - 3y + 4y^3 = 0$$

$$2x^2y\frac{dy}{dx} - 3x\frac{dy}{dx} + 12xy^2\frac{dy}{dx} = 3y - 2xy^2 - 4y^3$$

Factorizing the left hand side gives:

$$\frac{dy}{dx}(2x^2y - 3x + 12xy^2) = 3y - 2xy^2 - 4y^3$$

$$\frac{dy}{dx} = \frac{3y - 2xy^2 - 4y^3}{2x^2y - 3x + 12xy^2}$$

4. Differentiate $x^4 + 6x^2y^2 - 5 = 0$ implicitly with respect to x.

Solution

$$x^4 + 6x^2y^2 - 5 = 0$$

We differentiate accordingly and apply product rule to $6x^2y^2$ as follows:

$$4x^3 + (6x^2 \times 2y\frac{dy}{dx}) + (y^2 \times 12x) - 0 = 0$$

$$4x^3 + 12x^2y\frac{dy}{dx} + 12xy^2 = 0$$

$$12x^2y\frac{dy}{dx} = -4x^3 - 12xy^2$$

216

$$\frac{dy}{dx} = \frac{-4x^3 - 12xy^2}{12x^2y}$$

$$= \frac{-4x(x^2 + 3y^2)}{12x^2y}$$

$$\frac{dy}{dx} = \frac{-(x^2 + 3y^2)}{3xy} \qquad \text{(4 and } x \text{ cancels out)}$$

5. Find $\frac{dy}{dx}$ if $\frac{x^2}{16} + \frac{y^2}{25} = 1$

<u>Solution</u>

$$\frac{x^2}{16} + \frac{y^2}{25} = 1$$

$$\frac{2x}{16} + \frac{2y}{25}\frac{dy}{dx} = 0$$

$$\frac{2y}{25}\frac{dy}{dx} = -\frac{2x}{16}$$

$$\frac{2y}{25}\frac{dy}{dx} = -\frac{x}{8}$$

$$\frac{dy}{dx} = \frac{-\frac{x}{8}}{\frac{2y}{25}}$$

$$= -\frac{x}{8} \times \frac{25}{2y}$$

$$\frac{dy}{dx} = -\frac{25x}{16y}$$

Exercise 18

1. Differentiate implicitly, the expression: $5x^3 + 3y = y^2$.

2. If $2x^2 + 3y^3 = 10$, find $\frac{dy}{dx}$

3. Find $\frac{dy}{dx}$ given that $xy - 3xy^2 + 4x^3 = 4$

4. Differentiate $2x^2 + xy^2 - 5 = x^3$ implicitly with respect to x.

5. Find $\frac{dy}{dx}$ if $\frac{x^3}{3} - \frac{y^2}{2} = 0$

6. Differentiate implicitly, the expression: $2x^2y + 5y = 1$.

7. If $2x^2y + 4y^3 = 2y$, find $\frac{dy}{dx}$

8. Find $\frac{dy}{dx}$ given that $5y^2 - 4xy = y$

217

9. Differentiate $x + y^2 - x^2 y^2 = 7$ implicitly with respect to x.

10. Find $\dfrac{dy}{dx}$ if $x^5 - \dfrac{3}{y} = 2$

CHAPTER 19
DERIVATIVE OF TRIGONOMETRIC FUNCTIONS

The derivatives of trigonometric functions are as given below:

If y = sin$x$, then $\frac{dy}{dx}$ = cos$x$

If y = cos$x$, then $\frac{dy}{dx}$ = $-$ sin$x$

If y = tan$x$, then $\frac{dy}{dx}$ = sec^2x

If y = cot$x$, then $\frac{dy}{dx}$ = $-$ cosec^2x

If y = sec$x$, then $\frac{dy}{dx}$ = sec$x$tan$x$

If y = cosec$x$, then $\frac{dy}{dx}$ = $-$ cot$x$cosec$x$

Example

1. Find the derivative of cos5x

<u>Solution</u>

Let y = cos5x

We have to use chain rule from composite function due to 5x which is a function of x.

Hence, let u = 5x

Therefore, y = cosu   (By replacing 5x with u)

$\frac{du}{dx}$ = 5

$\frac{dy}{du}$ = $-$ sinu   (Recall that the derivative of cos$x$ is $-$sin$x$)

Therefore, $\frac{dy}{dx}$ = $\frac{dy}{du}$ x $\frac{du}{dx}$ (Chain rule)

= $-$ sinu x 5

= $-$ 5sinu

$\frac{dy}{dx}$ = $-$ 5sin5x (Since u = 5x)

2. If y = sin$\frac{1}{2}$$x$, find $\frac{dy}{dx}$

<u>Solution</u>

y = sin$\frac{1}{2}$$x$

Let u = $\frac{1}{2}$$x$

Therefore, y = sinu   (By replacing $\frac{1}{2}$$x$ with u)

$$\frac{du}{dx} = \frac{1}{2}$$

$$\frac{dy}{du} = \cos u \qquad \text{(Recall that the derivative of } \sin x \text{ is } \cos x\text{)}$$

Therefore, $\dfrac{dy}{dx} = \dfrac{dy}{du} \times \dfrac{du}{dx}$ (Chain rule)

$$= \cos u \times \frac{1}{2}$$

$$= \frac{1}{2}\cos u$$

$$\frac{dy}{dx} = \frac{1}{2}\cos\frac{1}{2}x \qquad \left(\text{Since } u = \frac{1}{2}x\right)$$

3. Find the derivative of $5\cos 3x$

Solution

    $y = 5\cos 3x$

Let $u = 3x$

Therefore, $y = 5\cos u$

$$\frac{du}{dx} = 3$$

$$\frac{dy}{du} = 5(-\sin u) \qquad \text{(Note that the constant term i.e. 5 should be used to multiply the derivative)}$$

$$= -5\sin u$$

Therefore, $\dfrac{dy}{dx} = \dfrac{dy}{du} \times \dfrac{du}{dx}$

$$= -5\sin u \times 3$$

$$= -15\sin u$$

$$\frac{dy}{dx} = -15\sin 3x \qquad \text{(Since } u = 3x\text{)}$$

4. Find the derivative of $\sin^2 x$

Solution

    $y = \sin^2 x$

Note that $\sin^2 x = \sin x \times \sin x$

Hence, let $u = \sin x$

Therefore, $y = u^2$ (i.e. $u \times u$ from $\sin x \times \sin x$)

$$\frac{du}{dx} = \cos x$$

$$\frac{dy}{du} = 2u$$

Therefore, $\dfrac{dy}{dx} = \dfrac{dy}{du} \times \dfrac{du}{dx}$

$$= 2u \times \cos x$$

$$= 2u\cos x$$

$$\frac{dy}{dx} = 2\sin x\cos x \qquad (\text{Since } u = \sin x)$$

5. Differentiate with respect to x: $y = \tan 2x$

<u>Solution</u>

$$y = \tan 2x$$

$$\frac{dy}{dx} = \sec^2 2x \times \frac{d(2x)}{dx} \qquad (\text{Note that the derivative of } \tan x \text{ is } \sec^2 x)$$

$$= \sec^2 2x \times 2$$

$$= 2\sec^2 2x$$

6. If $y = \cos^3 6x^2$, differentiate y with respect to x.

<u>Solution</u>

$$y = \cos^3 6x^2$$

Let us solve this problem without the use of v as follows:

Let $u = \cos 6x^2$

Hence, $y = u^3$

$$\frac{du}{dx} = 12x(-\sin 6x^2) \qquad (\text{Note that } 12x \text{ is from the derivative of } 6x^2)$$

$$= -12x\sin 6x^2$$

$$\frac{dy}{du} = 3u^2$$

Therefore, $\dfrac{dy}{dx} = \dfrac{dy}{du} \times \dfrac{du}{dx}$

$$= 3u^2 \times (-12x\sin 6x^2)$$

$$= -36xu^2\sin 6x^2$$

$$= -36x(\cos 6x^2)^2\sin 6x^2 \qquad (u \text{ has been replaced with } \cos 6x^2)$$

$$\frac{dy}{dx} = -36x\cos^2 6x^2\sin 6x^2$$

7. Find the derivative of $y = \sec 3x$

<u>Solution</u>

$$y = \sec 3x$$

$$\frac{dy}{dx} = \sec 3x\tan 3x \times \frac{d(3x)}{dx} \qquad (\text{Note that the derivative of } \sec x \text{ is } \sec x\tan x)$$

$$= \sec 3x\tan 3x \times 3$$

$$\frac{dy}{dx} = 3\sec 3x\tan 3x$$

8. Find the derivative of $\operatorname{cosec} 4x^3$

Solution

$$y = \text{cosec}4x^3$$

$$\frac{dy}{dx} = -\text{cosec}4x^3\cot4x^3 \times \frac{d(4x^3)}{dx}$$ (Note that the derivative of cosec is $-$cosec cot)

$$= -\text{cosec}4x^3\cot4x^3 \times 12x^2$$

$$= -12x^2\text{cosec}4x^3\cot4x^3$$

9. Find the derivative of $\cot^2 2x^4$

Solution

$$y = \cot^2 2x^4$$

This can also be written as: $y = \cot2x^4 \times \cot2x^4$

Let $u = 2x^4$ (Take the function of x)

Also, let $v = \cot u$ (Take a function of u without taking the exponent)

Hence, $y = v^2$ (Since $v^2 = (\cot u)^2 = (\cot2x^4)^2 = \cot^2 2x^4$. Hence, $y = v^2$)

$$\frac{du}{dx} = 8x^3$$

$$\frac{dv}{du} = -\text{cosec}^2 u$$

$$\frac{dy}{dv} = 2v$$

Therefore, $\dfrac{dy}{dx} = \dfrac{dy}{dv} \times \dfrac{dv}{du} \times \dfrac{du}{dx}$

$$= 2v \times -\text{cosec}^2 u \times 8x^3$$

$$= -16x^3 v \,\text{cosec}^2 u$$

$$= -16x^3 \cot u \,\text{cosec}^2 u$$ (Since $v = \cot u$)

Substituting in the original value of u gives:

$$\frac{dy}{dx} = -16x^3\cot 2x^4\text{cosec}^2 2x^4$$

10. Find the derivative of $5x\sin2x$

Solution

$$y = 5x\sin2x$$

We are going to apply the product rule of differentiation.

Let $u = 5x$

and $v = \sin2x$

$$\frac{du}{dx} = 5$$

$$\frac{dv}{dx} = 2\cos2x$$ (Note that the differentiation of $2x$ gives 2)

Hence, $\dfrac{dy}{dx} = u\dfrac{dv}{dx} + v\dfrac{du}{dx}$

$$= (5x \times 2\cos2x) + (\sin2x \times 5)$$

$$= 10x\cos 2x + 5\sin 2x$$

$$\frac{dy}{dx} = 5(2x\cos 2x + \sin 2x)$$

11. Find $\frac{dy}{dx}$ if $y = \frac{1}{x}\sec x$

Solution

$$y = \frac{1}{x}\sec x$$

We apply product rule as follows:

$$u = \frac{1}{x}$$

$$= x^{-1}$$

$$v = \sec 3x$$

$$\frac{du}{dx} = -1x^{-2}$$

$$= \frac{-1}{x^2}$$

$$\frac{dv}{dx} = \sec 3x\tan 3x \times 3$$

$$= 3\sec 3x\tan 3x$$

Note that the derivative of $\sec x$ is $\sec x\tan x$ and 3 is from the derivative of $3x$

Hence, $\frac{dy}{dx} = u\frac{dv}{dx} + v\frac{du}{dx}$

$$= \frac{1}{x}(3\sec 3x\tan 3x) + \sec 3x\left(\frac{-1}{x^2}\right)$$

$$\frac{dy}{dx} = \frac{3}{x}\sec 3x\tan 3x - \frac{1}{x^2}\sec 3x$$

12. Find the derivative of $3\cosec x^6$

Solution

$$y = 3\cosec x^6$$

Let $u = x^6$

Hence, $y = 3\cosec u$

$$\frac{du}{dx} = 6x^5$$

$$\frac{dy}{du} = -3\cot u \cosec u \quad \text{(The constant term i.e. 5 should be used to multiply the derivative)}$$

Therefore, $\frac{dy}{dx} = \frac{dy}{du} \times \frac{du}{dx}$

$$= -3\cot u \cosec u \times 6x^5$$

$$= -18x^5 \cot u \cosec u$$

$$\frac{dy}{dx} = -18x^5 \cot x^6 \cosec x^6 \quad \text{(Since } u = x^6\text{)}$$

13. If $y = \dfrac{1 + \cos 2x}{\sin 2x}$ find $\dfrac{dy}{dx}$.

Solution

$$y = \frac{1 + \cos 2x}{\sin 2x}$$

We have to apply quotient rule on this as follows:

$u = 1 + \cos 2x$

$v = \sin 2x$

$\dfrac{du}{dx} = -2\sin 2x$

$\dfrac{dv}{dx} = 2\cos 2x$

Hence, $\dfrac{dy}{dx} = \dfrac{v\frac{du}{dx} - u\frac{dv}{dx}}{v^2}$ (Quotient rule)

$$= \frac{\sin 2x(-2\sin 2x) - (1 + \cos 2x)(2\cos 2x)}{(\sin 2x)^2}$$

$$= \frac{-2\sin^2 2x - (2\cos 2x + 2\cos^2 2x)}{\sin^2 2x}$$ (Note that $(\sin 2x)^2 = \sin^2 2x$)

$$= \frac{-2\sin^2 2x - 2\cos 2x - 2\cos^2 2x}{\sin^2 2x}$$

$$= \frac{-2\sin^2 2x - 2\cos^2 2x - 2\cos 2x}{\sin^2 2x}$$

$$= \frac{-2(\sin^2 2x + \cos^2 2x) - 2\cos 2x}{\sin^2 2x}$$

$$= \frac{-2(1) - 2\cos 2x}{\sin^2 2x}$$ (Note that $\sin^2 x + \cos^2 x = 1$, hence $\sin^2 2x + \cos^2 2x = 1$)

$$\frac{dy}{dx} = \frac{-2(1 + \cos 2x)}{\sin^2 2x}$$

14. Differentiate $y = \dfrac{\sin^2 x}{x}$

Solution

$$y = \frac{\sin^2 x}{x}$$

This is also quotient rule.

Therefore, $u = \sin^2 x$

$v = x$

Let us follow a direct and systematic way of differentiating trigonometric functions.

Hence, in order to differentiate $\sin 2x$:

First, differentiate the exponent of $\sin^2$ without changing the trigonometric term. This gives:

$2\sin^{2-1}$

$= 2\sin$

224

Then add the term in x. This gives:

   $2\sin x$

The next step is to differentiate sin which gives cos. Also add the term in x to obtain $\cos x$.

Finally, differentiate the term in x. Hence, differentiating x gives 1.

Now multiply the three terms obtained in the three steps above. This gives:

   $2\sin x \; \times \; \cos x \; \times \; 1$

   $= 2\sin x \cos x$

Hence, $\dfrac{du}{dx} = 2\sin x\cos x$ (As obtained above)

Since, $v = x$

Then, $\dfrac{dv}{dx} = 1$

Hence, $\dfrac{dy}{dx} = \dfrac{v\frac{du}{dx} - u\frac{dv}{dx}}{v^2}$

$= \dfrac{x(2\sin x\cos x) - \sin^2 x(1)}{x^2}$

$= \dfrac{2x\sin x\cos x - \sin^2 x}{x^2}$

$= \dfrac{\sin x(2x\cos x - \sin x)}{x^2}$

15. Differentiate with respect to x: $\dfrac{\sec 2x}{x^3 + 1}$

Solution

   $y = \dfrac{\sec 2x}{x^3 + 1}$

   $u = \sec 2x$

   $v = x^3 + 1$

Hence, $\dfrac{du}{dx} = 2\sec 2x\tan 2x$

Then, $\dfrac{dv}{dx} = 3x^2$

Hence, $\dfrac{dy}{dx} = \dfrac{v\frac{du}{dx} - u\frac{dv}{dx}}{v^2}$

$= \dfrac{(x^3 + 1)(2\sec 2x\tan 2x) - \sec 2x(3x^2)}{(x^3 + 1)^2}$

$= \dfrac{(x^3 + 1)(2\sec 2x\tan 2x) - 3x^2\sec 2x}{(x^3 + 1)^2}$

16. Find the derivative of $\dfrac{\cos\sqrt{x}}{1 + x}$

<u>Solution</u>

$$y = \frac{\cos\sqrt{x}}{1+x}$$

$$u = \cos\sqrt{x}$$

$$= \cos x^{\frac{1}{2}}$$

$$v = 1 + x$$

Hence, $\dfrac{du}{dx} = \dfrac{d(x^{\frac{1}{2}})}{dx} \times \dfrac{d(\cos)}{dx}$

Note that $\dfrac{d(\cos)}{dx}$ means the derivative of cos which gives $-\sin$, and then $x^{\frac{1}{2}}$ is added to it to give $-\sin x^{\frac{1}{2}}$. Hence we continue as follows:

$$\frac{du}{dx} = \frac{1}{2} x^{-\frac{1}{2}} \times -\sin x^{\frac{1}{2}}$$

$$= \frac{1}{2x^{\frac{1}{2}}} \times -\sin x^{\frac{1}{2}}$$

$$\frac{du}{dx} = \frac{-\sin x^{\frac{1}{2}}}{2x^{\frac{1}{2}}}$$

Also, $\dfrac{dv}{dx} = 1$

Hence, $\dfrac{dy}{dx} = \dfrac{v\dfrac{du}{dx} - u\dfrac{dv}{dx}}{v^2}$

$$= \frac{(1+x)\left(\dfrac{-\sin x^{\frac{1}{2}}}{2x^{\frac{1}{2}}}\right) - \cos x^{\frac{1}{2}}(1)}{(1+x)^2}$$

$$= \frac{-\sin\sqrt{x}}{2\sqrt{x}(1+x)} - \frac{\cos\sqrt{x}}{(1+x)^2} \qquad \text{(When the fractions are separated)}$$

$$\frac{dy}{dx} = \frac{-(1+x)\sin\sqrt{x} - 2\sqrt{x}\cos\sqrt{x}}{2\sqrt{x}(1+x)^2} \qquad \text{(When the fractions are combined)}$$

17. Find $\dfrac{dy}{dx}$ if $y = \dfrac{1-x^2}{1+\cos x}$

<u>Solution</u>

$$y = \frac{1-x^2}{1+\cos x}$$

$$u = 1 - x^2$$

$$v = 1 + \cos x$$

$$\frac{du}{dx} = -2x$$

$$\frac{dv}{dx} = -\sin x$$

Hence, $\dfrac{dy}{dx} = \dfrac{v\frac{du}{dx} - u\frac{dv}{dx}}{v^2}$

$$= \dfrac{(1+\cos x)(-2x) - (1-x^2)(-\sin x)}{(1+\cos x)^2}$$

$$\dfrac{dy}{dx} = \dfrac{-2x(1+\cos x) + (1-x^2)\sin x}{(1+\cos x)^2}$$

Note that the negative sign from $-\sin x$ changed the negative sign at the middle to a positive sign since negative sign multiplied by negative sign gives a positive sign.

18. Find the derivative of $8x\sin x^2$

Solution

$\quad y = 8x\sin x^2$

We apply product rule as follows:

$\quad u = 8x$

$\quad \dfrac{du}{dx} = 8$

$\quad v = \sin x^2$

$\quad \dfrac{dv}{dx} = \dfrac{d(\sin)}{dx} \times \dfrac{d(x^2)}{dx}$

$\qquad = \cos x^2 \times 2x$

$\qquad = 2x\cos x^2$

Note that $\dfrac{d(\sin)}{dx}$ gives cos which result to $\cos x^2$ when x^2 from the question is added

Hence, $\dfrac{dy}{dx} = u\dfrac{dv}{dx} + v\dfrac{du}{dx}$

$\qquad = 8x(2x\cos x^2) + \sin x^2(8)$

$\qquad = 16x^2\cos x^2 + 8\sin x^2$

$\dfrac{dy}{dx} = 8(2x^2\cos x^2 + \sin x^2)$

19. Differentiate with respect to x: $\sec 2x\sin^3 2x$

Solution

$\quad y = \sec 2x\sin^3 2x$

We apply product rule as follows:

$\quad u = \sec 2x$

$\quad \dfrac{du}{dx} = \dfrac{d(\sec)}{dx} \times \dfrac{d(2x)}{dx}$

$\qquad = \sec 2x\tan 2x \times 2$

$\quad \dfrac{du}{dx} = 2\sec 2x\tan 2x$

227

Note that $\dfrac{d(\sec)}{dx}$ gives sectan, but remember to add $2x$ after sec and tan respectively to obtain

sec2xtan2x

$$v = \sin^3 2x$$

In order to directly differentiate a trigonometric term with exponent like this (i.e. $\sin^3$) we have three solutions to multiply as follows:

First solution: consider only the exponent and differentiate $\sin^3$. This gives:

$$\frac{d(\sin^3)}{dx} = 3\sin^2$$

We now add $2x$ from the question to obtain $3\sin^2 2x$

Second solution: $\dfrac{d(\sin)}{dx} = \cos$ which gives $\cos 2x$

Third solution: $\dfrac{d(2x)}{dx} = 2$

Multiply theses three solutions to give the derivative of $v = \sin^3 2x$ as follows:

$$\frac{dv}{dx} = 3\sin^2 2x \times \cos 2x \times 2$$

$$= 6\sin^2 2x \cos 2x$$

Now, $\dfrac{dy}{dx} = u\dfrac{dv}{dx} + v\dfrac{du}{dx}$

$$= \sec 2x(6\sin^2 2x \cos 2x) + \sin^3 2x(2\sec 2x \tan 2x)$$

$$= \frac{1}{\cos 2x}(6\sin^2 2x \cos 2x) + \sin^3 2x(2\frac{1}{\cos 2x}\frac{\sin 2x}{\cos 2x}) \quad \text{(Note that } \sec 2x = \frac{1}{\cos 2x} \text{ and } \tan 2x = \frac{\sin 2x}{\cos 2x})$$

$$= 6\sin^2 2x + 2\sin^2 2x(\frac{\sin 2x}{\cos 2x}\frac{\sin 2x}{\cos 2x})$$

Note that one sin2x has been taken out of $\sin^3 2x$ and placed inside the bracket.

$$= 6\sin^2 2x + 2\sin^2 2x(\tan 2x \tan x) \quad \text{(Since } \frac{\sin 2x}{\cos 2x} = \tan 2x)$$

$$= 6\sin^2 2x + 2\sin^2 2x(\tan^2 2x)$$

$$\frac{dy}{dx} = 2\sin^2 2x(3 + \tan^2 2x) \quad \text{(After factorization)}$$

20. Differentiate with respect to x: $\dfrac{\sec x - \tan x}{\sec x + \tan x}$

Solution

$$y = \frac{\sec x - \tan x}{\sec x + \tan x}$$

$$u = \sec x - \tan x$$

$$\frac{du}{dx} = \sec x \tan x - \sec^2 x$$

$$v = \sec x + \tan x$$

$$\frac{dv}{dx} = \sec x \tan x + \sec^2 x$$

$$\frac{dy}{dx} = \frac{v\frac{du}{dx} - u\frac{dv}{dx}}{v^2}$$

$$= \frac{(\sec x + \tan x)(\sec x\tan x - \sec 2x) - (\sec x - \tan x)(\sec x\tan x + \sec 2x)}{(\sec x + \tan x)^2}$$

Expanding bracket in the numerator gives:

$$\frac{dy}{dx} = \frac{\sec^2 x\tan x - \sec^3 x + \sec x\tan^2 x - \sec^2 x\tan x - [\sec^2 x\tan x + \sec^3 x - \sec x\tan^2 x - \sec^2 x\tan x]}{(\sec x + \tan x)^2}$$

$$= \frac{\sec^2 x\tan x - \sec^3 x + \sec x\tan^2 x - \sec^2 x\tan x - \sec^2 x\tan x - \sec^3 x + \sec x\tan^2 x + \sec^2 x\tan x]}{(\sec x + \tan x)^2}$$

$$= \frac{2\sec x\tan^2 x - 2\sec^3 x}{(\sec x + \tan x)^2} \qquad \text{(Note that all the sec}^2 x\text{tan}x \text{ have cancelled out)}$$

$$= \frac{2\sec x(\tan^2 x - \sec^2 x)}{(\sec x + \tan x)^2}$$

$$= \frac{2\sec x(-1)}{(\sec x + \tan x)^2} \qquad \text{(Note that tan}^2 x - \text{sec}^2 x = -1\text{)}$$

$$= \frac{-2\sec x}{(\sec x + \tan x)^2}$$

$$\frac{dy}{dx} = -\frac{2\sec x}{(\sec x + \tan x)^2}$$

21. Find the derivative of $\sqrt{\dfrac{\cos 2x}{1 + \sin 2x}}$

Solution

$$y = \sqrt{\frac{\cos 2x}{1 + \sin 2x}}$$

This can also be written as:

$$y = \frac{(\cos 2x)^{\frac{1}{2}}}{(1 + \sin 2x)^{\frac{1}{2}}}$$

$$u = (\cos 2x)^{\frac{1}{2}}$$

Using the chain rule, we obtain $\dfrac{du}{dx}$ as follows:

$$\frac{du}{dx} = \frac{1}{2}(\cos 2x)^{\frac{1}{2} - 1} \times \frac{d(\cos 2x)}{dx}$$

$$= \frac{1}{2}(\cos 2x)^{-\frac{1}{2}} \times -2\sin 2x$$

$$= \frac{-2\sin 2x}{2(\cos 2x)^{\frac{1}{2}}}$$

$$= \frac{-\sin 2x}{(\cos 2x)^{\frac{1}{2}}}$$

$$v = (1 + \sin 2x)^{\frac{1}{2}}$$

229

$$\frac{dv}{dx} = \frac{1}{2}(1 + \sin 2x)^{\frac{1}{2}-1} \times \frac{d(1+\sin 2x)}{dx}$$

$$= \frac{1}{2}(1 + \sin 2x)^{-\frac{1}{2}} \times 2\cos 2x$$

$$= \frac{2\cos 2x}{2(1 + \sin 2x)^{\frac{1}{2}}}$$

$$= \frac{\cos 2x}{(1 + \sin 2x)^{\frac{1}{2}}}$$

$$\frac{dy}{dx} = \frac{v\frac{du}{dx} - u\frac{dv}{dx}}{v^2}$$

$$= \frac{(1 + \sin 2x)^{\frac{1}{2}}\left(\frac{-\sin 2x}{(\cos 2x)^{\frac{1}{2}}}\right) - (\cos 2x)^{\frac{1}{2}}\left(\frac{\cos 2x}{(1 + \sin 2x)^{\frac{1}{2}}}\right)}{[(1 + \sin 2x)^{\frac{1}{2}}]^2}$$

$$= \frac{\left(\frac{-\sin 2x(1 + \sin 2x)^{\frac{1}{2}}}{(\cos 2x)^{\frac{1}{2}}}\right) - \left(\frac{\cos 2x(\cos 2x)^{\frac{1}{2}}}{(1 + \sin 2x)^{\frac{1}{2}}}\right)}{1 + \sin 2x}$$

$$= \frac{\left(\frac{-\sin 2x(1 + \sin 2x) - \cos 2x(\cos 2x)}{(\cos 2x)^{\frac{1}{2}}(1 + \sin 2x)^{\frac{1}{2}}}\right)}{1 + \sin 2x}$$

Note that $(\cos 2x)^{\frac{1}{2}} \times (\cos 2x)^{\frac{1}{2}} = \cos 2x$ (Since the exponents are added). Similarly:

$(1 + \sin 2x)^{\frac{1}{2}} \times (1 + \sin 2x)^{\frac{1}{2}} = 1 + \sin 2x$. Simplifying the above expression further, gives:

$$\frac{dy}{dx} = \frac{-\sin 2x(1 + \sin 2x) - \cos^2 2x}{(\cos 2x)^{\frac{1}{2}}(1 + \sin 2x)^{\frac{1}{2}}(1 + \sin 2x)}$$

$$= \frac{-\sin 2x - \sin^2 2x - \cos^2 2x}{(\cos 2x)^{\frac{1}{2}}(1 + \sin 2x)^{\frac{1}{2}}(1 + \sin 2x)}$$

$$= \frac{-\sin 2x - (\sin^2 2x + \cos^2 2x)}{(\cos 2x)^{\frac{1}{2}}(1 + \sin 2x)^{\frac{1}{2}}(1 + \sin 2x)}$$

$$= \frac{-\sin 2x - 1}{(\cos 2x)^{\frac{1}{2}}(1 + \sin 2x)^{\frac{1}{2}}(1 + \sin 2x)} \qquad \text{(Note that } \sin^2 2x + \cos^2 2x = 1)$$

$$= \frac{-(1 + \sin 2x)}{(\cos 2x)^{\frac{1}{2}}(1 + \sin 2x)^{\frac{1}{2}}(1 + \sin 2x)}$$

$$= \frac{-1}{(\cos 2x)^{\frac{1}{2}}(1 + \sin 2x)^{\frac{1}{2}}} \qquad \text{(Note that } 1 + \sin 2x \text{ cancels out)}$$

22. Differentiate with respect to x: $\sin x - 2x\cos x$

Solution

$$y = \sin x - 2x\cos x$$

Treat $2x\cos x$ using product rule.

$$\frac{dy}{dx} = \frac{d(\sin x)}{dx} - \left(2x\,\frac{d(\cos x)}{dx} + \cos x\,\frac{d(2x)}{dx}\right)$$

$$= \cos x - [2x(-\sin x) + \cos x(2)]$$

$$= \cos x + 2x\sin x - 2\cos x$$

$$= 2x\sin x - \cos x$$

23. Find the derivative of $\cos^5 3x^4$

Solution

$$y = \cos^5 3x^4$$

A direct way of differentiating this problem is applied as follows:

$$\frac{dy}{dx} = \frac{d(\cos^5)}{dx} \times \frac{d(\cos)}{dx} \times \frac{d(3x^4)}{dx}$$

$$= 5\cos^4 3x^4 \times (-\sin 3x^4) \times 12x^3$$

Note that the derivative of $\cos^5$ gives $5\cos^4$ (do not change the trigonometric term, i.e. cos) and then the addition of $3x^4$ from the question gives $5\cos^4 3x^4$.

Similarly the derivative of cos gives $-\sin$ and the addition of $3x^4$ from the question gives $-\sin 3x^4$.

Hence, multiplying the terms above gives:

$$\frac{dy}{dx} = -60x^3\cos^4 3x^4\sin 3x^4$$

24. Differentiate with respect to x: $\sin^8 15x^6$

Solution

$$y = \sin^8 15x^6$$

We can also differentiate this problem directly as follows:

$$\frac{dy}{dx} = \frac{d(\sin^8)}{dx} \times \frac{d(\sin)}{dx} \times \frac{d(15x^6)}{dx}$$

$$= 8\sin^7 15x^6 \times \cos 15x^6 \times 90x^5$$

Note that the derivative of $\sin^8$ gives $8\sin^7$ (in this case do not change the trigonometric term, i.e. sin) and then the addition of $15x^6$ from the question gives $8\sin^7 15x^6$.

Similarly the derivative of sin gives cos and the addition of $15x^6$ from the question gives $\cos 15x^6$. Hence, we continue as follows:

$$\frac{dy}{dx} = 8\sin^7 15x^6 \times \cos 15x^6 \times 90x^5$$

$$= 720x^5\sin^7 15x^6\cos 15x^6$$

Exercise 19

1. Find the derivative of $\tan 2x$

2. If $y = \cos\dfrac{1}{5}x$, find $\dfrac{dy}{dx}$

3. Find the derivative of $10\cos 5x$

4. Find the derivative of $\sin^3 x$

5. Differentiate with respect to x: $y = \tan^2 x$

6. If $y = \sin^4 3x^5$, differentiate y with respect to x.

7. Find the derivative of $y = \operatorname{cosec} 6x^2$

8. Find the derivative of $\sec 2x^5$

9. Find the derivative of $\tan 3x^2$

10. Find the derivative of $x^2\cos 3x$

11. Find $\dfrac{dy}{dx}$ if $y = \dfrac{3}{x^3}\sin 2x$

12. Find the derivative of $12\sec x^4$

13. If $y = \dfrac{2-\sin 5x}{\tan x}$ find $\dfrac{dy}{dx}$.

14. Differentiate $y = \dfrac{\sin^3 x}{2x}$

15. Differentiate with respect to x: $\dfrac{\cot 2x}{x+3}$

16. Find the derivative of $\dfrac{\sin\sqrt[3]{3}}{5x}$

17. Find $\dfrac{dy}{dx}$ if $y = \dfrac{x^2-3}{\sec 2x}$

18. Find the derivative of $3x^2\cos 3x^2$

19. Differentiate with respect to x: $\cot x\cos^2 x$

20. Differentiate with respect to x: $\dfrac{\sin x - \cos x}{\sin x + \cos x}$

21. Find the derivative of $\dfrac{\sec 4x}{\cos 4x - 2}$

22. Differentiate with respect to x: $\cos 3x - x^2\sec x$

23. Find the derivative of $\sin^9 x^3$

24. Differentiate with respect to x: $\sin 6x^{\frac{1}{2}}$

25. Find $\dfrac{dy}{dx}$ if $y = \dfrac{1}{x^2}\sin x^3$

26. Find the derivative of $\cos^3 2x^5$

27. If $y = \dfrac{\cos 10x}{\sin 2x}$ find $\dfrac{dy}{dx}$.

28. Differentiate $y = \dfrac{3\sin x^4}{2x}$

29. Differentiate with respect to x: $\dfrac{\tan 5x}{2x - 1}$

30. Find the derivative of $\dfrac{\tan^2 x}{2x}$

CHAPTER 20
DERIVATIVE OF INVERSE FUNCTIONS

If the derivative of a function is given by $\dfrac{dy}{dx}$, then the derivative of the inverse function is given

by: $\dfrac{1}{\dfrac{dy}{dx}} = \dfrac{dx}{dy}$

Or, $\dfrac{dy}{dx} = \dfrac{1}{\dfrac{dx}{dy}}$

Examples

1. Find $\dfrac{dx}{dy}$ if $y = \sqrt[3]{x}$

Solution

Method 1

$y = \sqrt[3]{x}$

Or $y = x^{\frac{1}{3}}$

Let us make x the subject of the formula. The inverse of $\dfrac{1}{3}$ is 3. Hence raise both sides to the

exponent 3 as follows:

$y^3 = (x^{\frac{1}{3}})^3$

$y^3 = x^1$ (Note that 1 was obtained from $\dfrac{1}{3} \times 3$)

Hence, $x = y^3$

Therefore, $\dfrac{dx}{dy} = 3y^2$

Method 2

$y = \sqrt[3]{x}$

Or $y = x^{\frac{1}{3}}$

$\dfrac{dy}{dx} = \dfrac{1}{3}x^{\frac{1}{3}-1}$

$= \dfrac{1}{3}x^{-\frac{2}{3}}$

$\dfrac{dy}{dx} = \dfrac{1}{3x^{\frac{2}{3}}}$

Hence, $\dfrac{dx}{dy} = \dfrac{1}{\dfrac{dy}{dx}}$

$= \dfrac{3x^{\frac{2}{3}}}{1}$ (This means the inverse of $\dfrac{1}{3x^{\frac{2}{3}}}$)

$= 3x^{\frac{2}{3}}$

$= 3(\sqrt[3]{x})^2$ [Recall from indices that $x^{\frac{a}{b}} = (\sqrt[b]{x})^a$]

$\dfrac{dx}{dy} = 3y^2$ (Since $y = \sqrt[3]{x}$)

2. If $y = \sqrt[5]{2x - 3}$ find $\dfrac{dx}{dy}$

Solution

$y = \sqrt[5]{2x - 3}$

Or $y = (2x - 3)^{\frac{1}{5}}$

Let us make x the subject of the formula. The inverse of $\dfrac{1}{5}$ is 5. Hence raise both sides to the exponent 5 as follows:

$y^5 = [(2x - 3)^{\frac{1}{5}}]^5$

$y^5 = 2x - 3$ (Note that $\dfrac{1}{5} \times 5 = 1$, which cancels the fractional exponent)

Hence, $2x = y^5 + 3$

$x = \dfrac{y^5 + 3}{2}$

$x = \dfrac{y^5}{2} + \dfrac{3}{2}$ (When we separate into fractions by dividing each part by the denominator)

Therefore, $\dfrac{dx}{dy} = \dfrac{5y^4}{2}$

3. If $y = x^3 - 5$. Find $\dfrac{dx}{dy}$

Solution

$y = x^3 - 5$

$y + 5 = x^3$

$x^3 = y + 5$

$x = (y + 5)^{\frac{1}{3}}$ (By raising both sides to an exponent of the inverse of 3 which is $\dfrac{1}{3}$)

$\dfrac{dx}{dy} = \dfrac{1}{3}(y + 5)^{\frac{1}{3} - 1} \times 1$

Note that the 1 was obtained from the derivative of y + 5 since chain rule was used.

$\dfrac{dx}{dy} = \dfrac{1}{3}(y + 5)^{-\frac{2}{3}}$

$\dfrac{dx}{dy} = \dfrac{1}{3(y+5)^{\frac{2}{3}}}$

Recall from indices that $x^{\frac{a}{b}} = (\sqrt[b]{x})^a$. Applying this rule gives:

$$\frac{dx}{dy} = \frac{1}{3[\sqrt[3]{(y+5)^2}]}$$

4. If $y = \frac{1}{2}x^4 + 3$, find $\frac{dx}{dy}$

Solution

$$y = \frac{1}{2}x^4 + 3$$

$$y - 3 = \frac{1}{2}x^4$$

$$2(y - 3) = x^4$$

$$x^4 = 2y - 6$$

$x = (2y - 6)^{\frac{1}{4}}$ (This is obtained by raising both sided to an exponent of the inverse of 4, i.e. $\frac{1}{4}$)

Hence we use chain rule to determine $\frac{dx}{dy}$ as follows:

$$\frac{dx}{dy} = \frac{1}{4}(2y - 6)^{\frac{1}{4} - 1} \times 2 \qquad \text{(Note that 2 is from the derivative of } 2y - 6)$$

$$\frac{dx}{dy} = \frac{1}{2}(2y - 6)^{-\frac{3}{4}}$$

$$\frac{dx}{dy} = \frac{1}{2(2y - 6)^{\frac{3}{4}}}$$

Hence, $\dfrac{dx}{dy} = \dfrac{1}{2[\sqrt[4]{(2y-6)^3}]}$ [This is obtained from the law of indices given by: $x^{\frac{a}{b}} = (\sqrt[b]{x})^a$]

5. If $y = \frac{x + 2}{x}$, find $\frac{dx}{dy}$.

Solution

$$y = \frac{x + 2}{x}$$

$xy = x + 2$ (When we cross multiply)

$xy - x = 2$

Factorizing the left hand side gives:

$x(y - 1) = 2$

$$x = \frac{2}{y - 1}$$

$x = 2(y - 1)^{-1}$ (Take note of the use of negative exponent when the denominator goes up)

Hence we use chain rule to determine $\frac{dx}{dy}$ as follows:

$$\frac{dx}{dy} = -1 \times 2(y - 1)^{-1-1} \times 1 \qquad \text{(Note that 1 is from the derivative of } y - 1)$$

$$\frac{dx}{dy} = -2(y-1)^{-2}$$

$$\frac{dx}{dy} = \frac{-2}{(y-1)^2}$$

6. Find $\frac{dx}{dy}$ if $y = \frac{1}{x+2}$

Solution

$$y = \frac{1}{x+2}$$

$$x + 2 = \frac{1}{y}$$

$$x = \frac{1}{y} - 2$$

$$x = y^{-1} - 2$$

$$\frac{dx}{dy} = -1 \times y^{-1-1}$$

$$= -y^{-2}$$

$$\frac{dx}{dy} = \frac{-1}{y^2}$$

7. Find $\frac{dx}{dy}$ if $y = x^{\frac{2}{3}}$

Solution

$$y = x^{\frac{2}{3}}$$

Raise both sides to the exponent $\frac{3}{2}$ i.e. the inverse of $\frac{2}{3}$. This gives:

$$y^{\frac{3}{2}} = (x^{\frac{2}{3}})^{\frac{3}{2}}$$

$$y^{\frac{3}{2}} = x \quad \text{(Note that } \frac{2}{3} \times \frac{3}{2} = 1, \text{ and } x^1 = x)$$

$$x = y^{\frac{3}{2}}$$

$$\frac{dx}{dy} = \frac{3}{2} \times y^{\frac{3}{2}-1}$$

$$= \frac{3}{2} y^{\frac{1}{2}}$$

$$\frac{dx}{dy} = \frac{3\sqrt{y}}{2}$$

8. If $y = (5x + 7)^{\frac{3}{10}}$ find $\frac{dx}{dy}$.

Solution

$$y = (5x + 7)^{\frac{3}{10}}$$

Raise both sides to the exponent $\frac{10}{3}$ i.e. the inverse of $\frac{3}{10}$. This gives:

$$y^{\frac{10}{3}} = [(5x + 7)^{\frac{3}{10}}]^{\frac{10}{3}}$$

$$y^{\frac{10}{3}} = 5x + 7 \quad \text{(Note that } \frac{3}{10} \times \frac{10}{3} = 1\text{)}$$

$$y^{\frac{10}{3}} - 7 = 5x$$

$$x = \frac{y^{\frac{10}{3}} - 7}{5}$$

$$= \frac{y^{\frac{10}{3}}}{5} - \frac{7}{5}$$

$$x = \frac{1}{5} y^{\frac{10}{3}} - \frac{7}{5}$$

$$\frac{dx}{dy} = \frac{10}{3} \times \frac{1}{5} y^{\frac{10}{3} - 1}$$

$$= \frac{2}{3} y^{\frac{7}{3}}$$

$$\frac{dx}{dy} = \frac{2\sqrt[3]{y^7}}{3}$$

Exercise 20

1. Find $\frac{dx}{dy}$ if $y = 7x^3$

2. If $y = \sqrt[3]{1 + x^2}$ find $\frac{dx}{dy}$

3. If $y = 2x^5 - 3$. Find $\frac{dx}{dy}$

4. If $y = \frac{2}{3}x^3 - 9$, find $\frac{dx}{dy}$

5. If $y = \frac{2x + 1}{x}$, find $\frac{dx}{dy}$.

6. Find $\frac{dx}{dy}$ if $y = \frac{3}{x^2 - 5}$

7. Find $\frac{dx}{dy}$ if $y = 2x^{\frac{1}{4}}$

8. If $y = (x - 3)^{\frac{1}{5}}$ find $\frac{dx}{dy}$.

9. If $y = \dfrac{4x + 5}{x}$, find $\dfrac{dx}{dy}$.

10. Find $\dfrac{dx}{dy}$ if $y = \dfrac{1}{x^3 + 8}$

CHAPTER 21
DERIVATIVES OF INVERSE TRIGONOMETRIC FUNCTIONS

Recall that if $\sin x = y$, then $x = \sin^{-1} y$. This is referred to as inverse trigonometric function. The derivatives of inverse trigonometric functions are given below.

If $y = \sin^{-1} x$, then $\dfrac{dy}{dx} = \dfrac{1}{\sqrt{1 - x^2}}$

If $y = \cos^{-1} x$, then $\dfrac{dy}{dx} = \dfrac{-1}{\sqrt{1 - x^2}}$

If $y = \tan^{-1} x$, then $\dfrac{dy}{dx} = \dfrac{1}{1 + x^2}$

If $y = \cot^{-1} x$, then $\dfrac{dy}{dx} = \dfrac{-1}{1 + x^2}$

If $y = \sec^{-1} x$, then $\dfrac{dy}{dx} = \dfrac{1}{x\sqrt{x^2 - 1}}$

If $y = \operatorname{cosec}^{-1} x$, then $\dfrac{dy}{dx} = \dfrac{-1}{x\sqrt{x^2 - 1}}$

Note that $\sin^{-1} x$ can also be written as $\arcsin x$. Other inverse function can be written in a similar way.

Examples

1. If $y = \cos^{-1} x$ find the $\dfrac{dy}{dx}$.

Solution

$\quad y = \cos^{-1} x$

$\quad \dfrac{dy}{dx} = \dfrac{-1}{\sqrt{1 - x^2}}$

2. Find the derivative of $\sin^{-1} 3x$

Solution

$\quad y = \sin^{-1} 3x$

Let us use the chain rule to solve this problem

Let $u = 3x$

Hence, $y = \sin^{-1} u$

$\quad \dfrac{du}{dx} = 3$

$\quad \dfrac{dy}{du} = \dfrac{1}{\sqrt{1 - u^2}}$

Therefore, $\dfrac{dy}{dx} = \dfrac{dy}{du} \times \dfrac{du}{dx}$ (Chain rule)

$$= \frac{1}{\sqrt{1-u^2}} \times 3$$

$$= \frac{3}{\sqrt{1-u^2}}$$

$$= \frac{3}{\sqrt{1-(3x)^2}} \qquad \text{(Since } u = 3x)$$

$$\frac{dy}{dx} = \frac{3}{\sqrt{1-9x^2}}$$

3. Find the derivative of $\cot^{-1}x^2$

Solution

$$y = \cot^{-1}x^2$$

Let $u = x^2$

Hence, $y = \cot^{-1}u$

$$\frac{du}{dx} = 2x$$

$$\frac{dy}{du} = \frac{-1}{1+u^2}$$

Therefore, $\dfrac{dy}{dx} = \dfrac{dy}{du} \times \dfrac{du}{dx}$

$$= \frac{-1}{1+u^2} \times 2x$$

$$= \frac{-2x}{1+u^2}$$

$$= \frac{-2x}{1+(x^2)^2} \qquad \text{(since } u = x^2)$$

$$\frac{dy}{dx} = \frac{-2x}{1+x^4}$$

4. If $y = \sec^{-1}2x^3$, find $y = \dfrac{dy}{dx}$

Solution

$$y = \sec^{-1}2x^3$$

Let $u = 2x^3$

Hence, $y = \sec^{-1}u$

$$\frac{du}{dx} = 6x^2$$

$$\frac{dy}{du} = \frac{1}{u\sqrt{u^2-1}}$$

Therefore, $\dfrac{dy}{dx} = \dfrac{dy}{du} \times \dfrac{du}{dx}$

$$= \frac{1}{u\sqrt{u^2-1}} \times 6x^2$$

$$= \frac{6x^2}{u\sqrt{u^2-1}}$$

$$= \frac{6x^2}{2x^3\sqrt{(2x^3)^2-1}} \quad \text{(since } u = 2x^3\text{)}$$

$$\frac{dy}{dx} = \frac{3}{x\sqrt{4x^6-1}} \quad \text{(Note that } \frac{6x^2}{2x^3} = \frac{3}{x}\text{)}$$

5. Find the derivative $x^2\tan^{-1}x$

Solution

$$y = x^2\tan^{-1}x$$

we apply the product rule as follows:

$$u = x^2$$

$$v = \tan^{-1}x$$

$$\frac{du}{dx} = 2x$$

$$\frac{dv}{dx} = \frac{1}{1+x^2}$$

$$\frac{dy}{dx} = u\frac{dv}{dx} + v\frac{du}{dx} \quad \text{(product rule)}$$

$$= x^2\left(\frac{1}{1+x^2}\right) + \tan^{-1}x(2x)$$

$$\frac{dy}{dx} = \frac{x^2}{1+x^2} + 2x\tan^{-1}x$$

6. If $y = 3x - 1 \cosec^{-1}x^3$, find $\frac{dy}{dx}$.

Solution

$$y = 3x - 1 \cosec^{-1}x^3$$

$$u = 3x - 1$$

$$\frac{du}{dx} = 3$$

$$v = \cosec^{-1}x^3$$

$$\frac{dv}{dx} = \frac{-1}{x^3\sqrt{(x^3)^2-1}} \times 3x^2 \quad \text{(From use of chain rule. Note that } 3x^2 \text{ is from the derivative of } x^3\text{)}$$

$$= \frac{-3x^2}{x^3\sqrt{x^6-1}}$$

$$\frac{dv}{dx} = \frac{-3}{x\sqrt{x^6-1}}$$

$$\frac{dy}{dx} = u\frac{dv}{dx} + v\frac{du}{dx} \quad \text{(product rule)}$$

$$= 3x - 1 \left(\frac{-3}{x\sqrt{x^6 - 1}}\right) + \cosec^{-1}x^3(3)$$

$$\frac{dy}{dx} = \frac{-3(3x - 1)}{x\sqrt{x^6 - 1}} + 3\cosec^{-1}x^3$$

7. Find the derivative of $\cos^{-1}(5x - 3)$

Solution

$$y = \cos^{-1}(5x - 3)$$

Let $u = 5x - 3$

Hence, $y = \cos^{-1}u$

$$\frac{du}{dx} = 5$$

$$\frac{dy}{du} = \frac{-1}{\sqrt{1 - u^2}}$$

Therefore, $\dfrac{dy}{dx} = \dfrac{dy}{du} \times \dfrac{du}{dx}$

$$= \frac{-1}{\sqrt{1 - u^2}} \times 5$$

$$= \frac{-5}{\sqrt{1 - u^2}}$$

$$= \frac{-5}{\sqrt{1 - (5x - 3)^2}} \qquad \text{(Since } u = 5x - 3)$$

Expanding the bracket gives:

$$\frac{dy}{dx} = \frac{-5}{\sqrt{1 - (25x^2 - 15x - 15x + 9)}}$$

$$= \frac{-5}{\sqrt{1 - 25x^2 + 15x + 15x - 9}}$$

$$\frac{dy}{dx} = \frac{-5}{\sqrt{-25x^2 + 30x - 8}}$$

8. If $y = \tan^{-1}\left(\dfrac{1}{m}\right)$, find $\dfrac{dy}{dm}$

Solution

$$y = \tan^{-1}\left(\frac{1}{m}\right)$$

Let $u = \dfrac{1}{m} = m^{-1}$

Hence, $y = \tan^{-1}u$

$$\frac{du}{dm} = -1m^{-2}$$

$$\frac{du}{dm} = \frac{-1}{m^2}$$

$$\frac{dy}{du} = \frac{1}{1 + u^2}$$

Therefore, $\dfrac{dy}{dm} = \dfrac{dy}{du} \times \dfrac{du}{dm}$

$= \dfrac{1}{1+u^2} \times \dfrac{-1}{m^2}$

$= \dfrac{1}{1+\left(\frac{1}{m}\right)^2} \times \dfrac{-1}{m^2}$

$= \dfrac{-1}{m^2\left[1+\left(\frac{1}{m}\right)^2\right]}$

$= \dfrac{-1}{m^2\left(1+\frac{1}{m^2}\right)}$

$\dfrac{dy}{dm} = \dfrac{-1}{m^2+1}$

9. If $y = \sin 3x$, find $\dfrac{dx}{dy}$.

Solution

$y = \sin 3x$

$\sin^{-1}y = 3x$ (Recall that if $\sin a = b$, then $a = \sin^{-1}b$)

$x = \dfrac{\sin^{-1}y}{3}$

$x = \dfrac{1}{3}\sin^{-1}y$

Hence, $\dfrac{dx}{dy} = \dfrac{1}{3}\dfrac{1}{\sqrt{1-y^2}}$

$\dfrac{dx}{dy} = \dfrac{1}{3\sqrt{1-y^2}}$

Note that this example asked us to find $\dfrac{dx}{dy}$ and not $\dfrac{dy}{dx}$

10. If $y = (\cos 5x)^2$ find $\dfrac{dx}{dy}$.

Solution

$y = (\cos 5x)^2$

This can also be written as $\cos^2 5x$. Hence:

$y = \cos^2 5x$.

Or, $\cos^2 5x = y$

taking the square root of both sides gives:

$\cos 5x = \sqrt{y}$

$5x = \cos^{-1}\sqrt{y}$

$$x = \frac{\cos^{-1}\sqrt{y}}{5}$$

$$= \frac{1}{5}\cos^{-1}\sqrt{y}$$

$$x = \frac{1}{5}\cos^{-1}y^{\frac{1}{2}}$$

Hence, $\dfrac{dx}{dy} = \dfrac{1}{5} \cdot \dfrac{-1}{\sqrt{1-(y^{\frac{1}{2}})^2}} \times \dfrac{1}{2}y^{-\frac{1}{2}}$ (Note that $\dfrac{1}{2}y^{-\frac{1}{2}}$ is from the derivative of $y^{\frac{1}{2}}$)

$$= \frac{-1}{5\sqrt{1-y}} \times \frac{1}{2y^{\frac{1}{2}}}$$

$$= \frac{-1}{10y^{\frac{1}{2}}\sqrt{1-y}}$$

$$= \frac{-1}{10\sqrt{y}\sqrt{1-y}}$$

$$\frac{dx}{dy} = \frac{-1}{10\sqrt{y(1-y)}}$$

Exercise 21

1. If $y = \sin^{-1}2x$ find the $\dfrac{dy}{dx}$.

2. Find the derivative of $\cos^{-1}x$

3. Find the derivative of $\cot^{-1}3x^2$

4. If $y = \cot^{-1}x^4$, find $y = \dfrac{dy}{dx}$

5. Find the derivative $3x\sec^{-1}x$

6. If $y = x^2 + 3\tan^{-1}x$, find $\dfrac{dy}{dx}$.

7. Find the derivative of $\tan^{-1}(x + 5)$

8. If $y = \cos^{-1}\left(\dfrac{1}{x}\right)$, find $\dfrac{dy}{dx}$

9. If $y = \cos 5x$, find $\dfrac{dx}{dy}$.

10. If $y = (\sec x)^2$ find $\dfrac{dx}{dy}$.

11. Find the derivative $5x\tan^{-1}3x$

12. If $y = x^2 + 4\sin^{-1}x^5$, find $\dfrac{dy}{dx}$.

13. Find the derivative of $\csc^{-1}x^3$

245

14. If $y = \tan 2x$, find $\dfrac{\mathrm{d}x}{\mathrm{d}y}$.

15. If $y = \sin^3 5x$, find $\dfrac{\mathrm{d}x}{\mathrm{d}y}$.

CHAPTER 22
DERIVATIVES OF HYPERBOLIC FUNCTIONS

Hyperbolic functions are functions in calculus which are expressed by the combination of the exponential functions e^x and e^{-x}. The derivatives of the six main hyperbolic functions are as given below.

1. If $y = \sinh x$, then $\dfrac{dy}{dx} = \cosh x$

2. If $y = \cosh x$, then $\dfrac{dy}{dx} = \sinh x$

3. If $y = \tanh x$, then $\dfrac{dy}{dx} = \operatorname{sech}^2 x$

4. If $y = \coth x$, then $\dfrac{dy}{dx} = -\operatorname{cosech}^2 x$

5. If $y = \operatorname{sech} x$, then $\dfrac{dy}{dx} = -\operatorname{sech} x \tanh x$

6. If $y = \operatorname{cosech} x$, then $\dfrac{dy}{dx} = -\operatorname{cosech} x \coth x$

Note that $\sinh x = \dfrac{e^x - e^{-x}}{2}$ and $\cosh x = \dfrac{e^x + e^{-x}}{2}$

Examples

1. If $y = \cosh x - 5\sinh x$, find $\dfrac{dy}{dx}$.

Solution

$$y = \cosh x - 5\sinh x$$

$$\frac{dy}{dx} = \sinh x - 5\cosh x$$

2. Find the derivative of $2x^3 \coth x$

Solution

$$y = 2x^3 \coth x$$

Using product rule gives $\dfrac{dy}{dx}$ as follows:

$$\frac{dy}{dx} = 2x^3 \left[\frac{d(\coth x)}{dx}\right] + \coth x \left[\frac{d(2x^3)}{dx}\right]$$

$$= 2x^3(-\operatorname{cosech}^2 x) + \coth x(6x^2)$$

$$= -2x^3 \operatorname{cosech}^2 x + 6x^2 \coth x$$

$$\frac{dy}{dx} = -2x^2(x\operatorname{cosech}^2 x - 3\coth x)$$

Take note of the change in sign of the term in the bracket. This is due to the negative sign outside the bracket. Expanding the bracket gives the original expression that was factorized.

3. If $y = \dfrac{\cosh x}{x^2 + 1}$ find $\dfrac{dy}{dx}$.

Solution

Let us use product rule to obtained $\dfrac{dy}{dx}$ as follows:

$$\frac{dy}{dx} = \frac{(x^2+1)\left[\frac{d(\cosh x)}{dx}\right] - \cosh x\left[\frac{d(x^2+1)}{dx}\right]}{(x^2+1)^2}$$

$$= \frac{(x^2+1)(\sinh x) - \cosh x(2x)}{(x^2+1)^2}$$

$$\frac{dy}{dx} = \frac{\sinh x(x^2+1) - 2x\cosh x}{(x^2+1)^2}$$

4. Find the derivative of $(\sinh 3x)^2$

Solution

$\quad$ $y = (\sinh 3x)^2$ $\quad$ (This can also be written as $\sinh^2 3x$)

Let $u = \sinh 3x$

Hence, $y = u^2$

$\quad$ $\dfrac{du}{dx} = 3\cosh 3x$ $\quad$ (Note that 3 is from the derivative of $3x$)

$\quad$ $\dfrac{dy}{du} = 2u$

$\quad$ $\dfrac{dy}{dx} = \dfrac{dy}{du}$ x $\dfrac{du}{dx}$

$\quad\quad$ $= 2u$ x $3\cosh 3x$

$\quad\quad$ $= 6u\cosh 3x$

$\quad\quad$ $= 6\sinh 3x \cosh 3x$ $\quad$ (u has been replaced with $\sinh 3x$)

5. Find the derivative of $\sinh^4 2x^3$

Solution

$\quad$ $y = \sinh^4 2x^3$

Let $u = \sinh 2x^3$

Hence, $y = u^4$

$\quad$ $\dfrac{du}{dx} = 6x^2\cosh 2x^3$ $\quad$ (Note that $6x^2$ is from the derivative of $2x^3$)

$\quad$ $\dfrac{dy}{du} = 4u^3$

$\quad$ $\dfrac{dy}{dx} = \dfrac{dy}{du}$ x $\dfrac{du}{dx}$

$\quad\quad$ $= 4u^3$ x $6x^2\cosh 2x^3$

$\quad\quad$ $= 24x^2u^3\cosh 2x^3$

$\quad$ $\dfrac{dy}{dx} = 24x^2\sinh^3 2x^3\cosh 2x^3$ $\quad$ (u has been replaced with $\sinh 2x^3$)

6. Find the derivative of $\text{cosech}4x^3$

Solution

$$y = \text{cosech}4x^3$$

$$\frac{dy}{dx} = -\text{cosech}4x^3\text{coth}4x^3 \times \frac{d(4x^3)}{dx}$$ (Note that the derivative of cosech is $-\text{cosechcoth}$)

$$= -\text{cosech}4x^3\text{coth}4x^3 \times 12x^2$$

$$= -12x^2\text{cosech}4x^3\text{coth}4x^3$$

7. Find the derivative of $\text{coth}^2 2x^4$

Solution

$$y = \text{coth}^2 2x^4$$

Let $u = \text{coth}2x^4$

Hence, $y = u^2$

$$\frac{du}{dx} = -\text{cosech}^2 2x^4 \times \frac{d(2x^4)}{dx}$$

$$= -\text{cosech}^2 2x^4 \times 8x^3$$

$$\frac{du}{dx} = -8x^3\text{cosech}^2 2x^4$$

$$\frac{dy}{du} = 2u$$

Therefore, $\dfrac{dy}{dx} = \dfrac{dy}{du} \times \dfrac{du}{dx}$

$$= 2u \times (-8x^3\text{cosech}^2 2x^4)$$

$$= -16x^3 u \, \text{cosech}^2 2x^4$$

$$\frac{dy}{dx} = -16x^3\text{coth}2x^4\text{cosech}^2 2x^4$$ (Since $u = \text{coth}2x^4$)

8. Find the derivative of $5x\text{sinh}2x$

Solution

$$y = 5x\text{sinh}2x$$

We apply the product rule of differentiation as follows:

Let $u = 5x$

and $v = \text{sinh}2x$

$$\frac{du}{dx} = 5$$

$$\frac{dv}{dx} = 2\text{cosh}2x$$ (Note that the differentiation of $2x$ gives 2)

Hence, $\dfrac{dy}{dx} = u\dfrac{dv}{dx} + v\dfrac{du}{dx}$

$$= (5x \times 2\text{cosh}2x) + (\text{sinh}2x \times 5)$$

$$= 10x\text{cosh}2x + 5\text{sinh}2x$$

$$\frac{dy}{dx} = 5(2x\cosh 2x + \sinh 2x)$$

9. If $y = \dfrac{1 + \cosh 2x}{\sinh 2x}$ find $\dfrac{dy}{dx}$.

Solution

$$y = \frac{1 + \cosh 2x}{\sinh 2x}$$

We apply quotient rule as follows:

$u = 1 + \cosh 2x$

$v = \sinh 2x$

$$\frac{du}{dx} = 2\sinh 2x$$

$$\frac{dv}{dx} = 2\cosh 2x$$

Hence, $\dfrac{dy}{dx} = \dfrac{v\frac{du}{dx} - u\frac{dv}{dx}}{v^2}$ (Quotient rule)

$$= \frac{\sinh 2x(2\sinh 2x) - (1 + \cosh 2x)(2\cosh 2x)}{(\sinh 2x)^2}$$

$$= \frac{2\sin^2 2x - (2\cosh 2x + 2\cosh^2 2x)}{\sinh^2 2x}$$ (Note that $(\sinh 2x)^2 = \sinh^2 2x$)

$$= \frac{2\sinh^2 2x - 2\cosh 2x - 2\cosh^2 2x}{\sinh^2 2x}$$

$$= \frac{2\sinh^2 2x - 2\cosh^2 2x - 2\cosh 2x}{\sinh^2 2x}$$

$$= \frac{-2(\cosh^2 2x - \sinh^2 2x) - 2\cosh 2x}{\sinh^2 2x}$$

$$= \frac{-2(1) - 2\cosh 2x}{\sinh^2 2x}$$ (Note that $\cosh^2 2x - \sinh^2 2x = 1$)

$$\frac{dy}{dx} = \frac{-2(1 + \cosh 2x)}{\sinh^2 2x}$$

Or, $\dfrac{dy}{dx} = \dfrac{-2(1 + \cosh 2x)}{\cosh^2 2x - 1}$ (Note that since $\cosh^2 2x - \sinh^2 2x = 1$, then $\cosh^2 2x - 1 = \sinh^2 2x$)

$$= \frac{-2(1 + \cosh 2x)}{(\cosh 2x + 1)(\cosh 2x - 1)}$$

Note that from difference of two squares we have: $a^2 - b^2 = (a + b)(a - b)$.

Hence, $\cosh^2 2x - 1$ is also a difference of two squares since 1 is also 1^2. Therefore, $\cosh^2 2x - 1 = (\cosh 2x + 1)(\cosh 2x - 1)$ as represented above. Hence the expression above becomes:

$$\frac{dy}{dx} = \frac{-2}{(\cosh 2x - 1)}$$

10. Differentiate $y = \dfrac{\sinh^2 x}{x}$

Solution

$$y = \frac{\sinh^2 x}{x}$$

We use quotient rule as follows:

Therefore, $u = \sinh^2 x$

$\quad\quad v = x$

Hence, $\dfrac{du}{dx} = 2\sinh x \cosh x$

Since, $v = x$

Then, $\dfrac{dv}{dx} = 1$

Hence, $\dfrac{dy}{dx} = \dfrac{v\frac{du}{dx} - u\frac{dv}{dx}}{v^2}$

$$= \frac{x(2\sinh x \cosh x) - \sinh^2 x(1)}{x^2}$$

$$= \frac{2x\sinh x \cosh x - \sin^2 x}{x^2}$$

$$= \frac{\sinh x(2x\cosh x - \sinh x)}{x^2}$$

Exercise 22

1. If $y = \sinh 3x - 2\cosh x$, find $\dfrac{dy}{dx}$.

2. Find the derivative of $5x^2 \operatorname{sech} x$

3. If $y = \dfrac{\sinh 2x}{3x^2}$ find $\dfrac{dy}{dx}$.

4. Find the derivative of $\cosh^2 3x$

5. Find the derivative of $\cosh^3 4x^5$

6. Find the derivative of $\coth 2x^5$

7. Find the derivative of $(\operatorname{sech} 2x^2)^3$

8. Find the derivative of $x^2 \cosh 5x$

9. If $y = \dfrac{\tanh 5x}{\coth 3x}$ find $\dfrac{dy}{dx}$.

10. Differentiate $y = \dfrac{2\cosh^3 x}{3x}$

CHAPTER 23
DERIVATIVE OF LOGARITHMIC FUNCTIONS

If $y = \log_a x$, then $\dfrac{dy}{dx} = \dfrac{1}{x}\log_a e$, where a is any base.

If $y = \log_e x$, then $\dfrac{dy}{dx} = \dfrac{1}{x}$

Note that $\log_e x$ can also be represented as $\ln x$ and the value of e is 2.718 (to 3 decimal places)

Examples

1. Find the derivative of $\log_a 2x$.

Solution

$$y = \log_a 2x$$

We will use chain rule since we have a function that is not just x but $2x$.

Let $u = 2x$

Therefore, $y = \log_a u$

$$\frac{du}{dx} = 2$$

$$\frac{dy}{du} = \frac{1}{u}\log_a e$$

$$\frac{dy}{dx} = \frac{dy}{du} \times \frac{du}{dx}$$

$$= \frac{1}{u}\log_a e \times 2$$

$$= \frac{2}{u}\log_a e$$

$$= \frac{2}{2x}\log_a e \qquad (\text{since } u = 2x)$$

$$\frac{dy}{dx} = \frac{1}{x}\log_a e$$

2. Find the derivative of $\log_a(5x - 1)$

Solution

$$y = \log_a(5x - 1)$$

Let $u = 5x - 1$

Therefore, $y = \log_a u$

$$\frac{du}{dx} = 5$$

$$\frac{dy}{du} = \frac{1}{u}\log_a e$$

$$\frac{dy}{dx} = \frac{dy}{du} \times \frac{du}{dx}$$

$$= \frac{1}{u} \log_a e \times 5$$

$$= \frac{5}{u} \log_a e$$

$$\frac{dy}{dx} = \frac{5}{5x-1} \log_a e \qquad (\text{since } u = 5x - 1)$$

3. Differentiate $\log_a(4x - 3)^2$ with respect to x

Solution

$$y = \log_a(4x - 3)^2$$

Let $u = (4x - 3)^2$

Therefore, $y = \log_a u$

$$\frac{du}{dx} = 2(4x - 3)^{2-1} \times 4 \qquad (\text{Note that 4 is from the derivative of } 4x - 3)$$

$$\frac{du}{dx} = 8(4x - 3)$$

$$\frac{dy}{du} = \frac{1}{u} \log_a e$$

$$\frac{dy}{dx} = \frac{dy}{du} \times \frac{du}{dx}$$

$$= \frac{1}{u} \log_a e \times 8(4x - 3)$$

$$= \frac{8(4x-3)}{u} \log_a e$$

$$= \frac{8(4x-3)}{(4x - 3)^2} \log_a e \qquad [\text{since } u = (4x - 3)^2]$$

$$\frac{dy}{dx} = \frac{8}{4x - 3} \log_a e \qquad (\text{One } 4x - 3 \text{ cancels out})$$

4. If $y = \log_a\sqrt{1 + 2x}$, find $\dfrac{dy}{dx}$

Solution

$$y = \log_a\sqrt{1 + 2x}$$

Let $u = \sqrt{1 + 2x} = (1 + 2x)^{\frac{1}{2}}$

Therefore, $y = \log_a u$

$$\frac{du}{dx} = \frac{1}{2}(1 + 2x)^{\frac{1}{2}-1} \times 2$$

$$= (1 + 2x)^{-\frac{1}{2}} \qquad (\text{Note that } \frac{1}{2} \times 2 = 1)$$

$$\frac{du}{dx} = \frac{1}{(1+2x)^{\frac{1}{2}}}$$

$$\frac{dy}{du} = \frac{1}{u} \log_a e$$

$$\frac{dy}{dx} = \frac{dy}{du} \times \frac{du}{dx}$$

$$= \frac{1}{u} \log_a e \times \frac{1}{(1+2x)^{\frac{1}{2}}}$$

$$= \frac{1}{(1+2x)^{\frac{1}{2}}} \log_a e \times \frac{1}{(1+2x)^{\frac{1}{2}}} \qquad \text{[Note that u has been replaced with } (1+2x)^{\frac{1}{2}}]$$

$$\frac{dy}{dx} = \frac{1}{1+2x} \log_a e \qquad \text{[Note that } (1+2x)^{\frac{1}{2}} \times (1+2x)^{\frac{1}{2}} = (1+2x)^1 \text{ by adding exponents]}$$

5. Find $\frac{dy}{dx}$ given that y = $\log_a \frac{1-3x}{1+3x}$

Solution

$$y = \log_a \frac{1-3x}{1+3x}$$

Or, y = $\log_a(1-3x) - \log_a(1+3x)$ $\qquad$ (Recall that $\log_x\left(\frac{a}{b}\right) = \log_x a - \log_x b$)

$$\frac{dy}{dx} = \frac{-3}{1-3x} \log_a e - \frac{3}{1+3x} \log_a e$$

$$= -\log_a e \left(\frac{3}{1-3x} + \frac{3}{1+3x}\right)$$

$$= -\log_a e \left[\frac{3(1+3x) + 3(1-3x)}{(1-3x)(1+3x)}\right]$$

$$= -\log_a e \left[\frac{3 + 9x + 3 - 9x}{1 - 9x^2}\right] \qquad \text{[Note that } (1-3x)(1+3x) = 1 - 9x^2]$$

$$= -\log_a e \left(\frac{6}{1 - 9x^2}\right)$$

$$\frac{dy}{dx} = \frac{-6}{1 - 9x^2} \log_a e$$

6. If y = $\log_{10}(x^2 - 2)$, find $\frac{dy}{dx}$.

Solution

$$y = \log_{10}(x^2 - 2)$$

The value 10 represents a in other examples. So we are going to differentiate y like the examples above except that we will write 10 wherever 'a' should be.

$$y = \log_{10}(x^2 - 2)$$

$$u = x^2 - 2$$

Hence, y = $\log_{10} u$

$$\frac{du}{dx} = 2x$$

$$\frac{dy}{du} = \frac{1}{u} \log_{10} e \qquad \text{(Just like differentiating } \log_a u)$$

$$\frac{dy}{dx} = \frac{dy}{du} \times \frac{du}{dx}$$

$$= \frac{1}{u} \log_{10}e \times 2x$$

$$= \frac{2x}{u} \log_{10}e$$

$$\frac{dy}{dx} = \frac{2x}{x^2 - 2} \log_{10}e$$

7. Find $\frac{dy}{dx}$ if $y = \log_{10} \frac{1}{x}$

Solution

$$y = \log_{10} \frac{1}{x}$$

Let us solve this question directly without using $u = \frac{1}{x}$ as follows:

$$\frac{dy}{dx} = \frac{\frac{d\left(\frac{1}{x}\right)}{dx}}{\frac{1}{x}} \times \log_{10}e$$

$$= \frac{\frac{d(x^{-1})}{dx}}{x^{-1}} \times \log_{10}e$$

$$= \frac{-x^{-2}}{x^{-1}} \times \log_{10}e \qquad \text{(Note that the derivative of } x^{-1} \text{ is } -x^{-2})$$

$$= -x^{-2} \times x^1 \times \log_{10}e \qquad \left(\text{Since } \frac{1}{x^{-1}} = x^1\right)$$

$$= -x^{-1} \times \log_{10}e \qquad \text{(After adding the exponents of } x)$$

$$\frac{dy}{dx} = -\frac{1}{x} \log_{10}e$$

8. Find the derivative of $\log_e(2 - x^3)$

Solution

$$y = \log_e(2 - x^3)$$

Note that the base here is 'e' and not 'a'.

Let $u = 2 - x^3$

Hence, $y = \log_e u$

$$\frac{du}{dx} = -3x^2$$

$$\frac{dy}{du} = \frac{1}{u} \qquad \left(\text{Recall that the derivative of } \log_e x = \frac{1}{x}\right)$$

$$\frac{dy}{dx} = \frac{dy}{du} \times \frac{du}{dx}$$

$$= \frac{1}{u} \times -3x^2$$

$$= \frac{-3x^2}{u}$$

$$\frac{dy}{dx} = \frac{-3x^2}{2 - x^3}$$

9. Find the derivative of $(\log_e 5x)^2$

Solution

$$y = (\log_e 5x)^2$$

Let $u = \log_e 5x$

Hence, $y = u^2$

$$\frac{du}{dx} = \frac{\frac{d(5x)}{dx}}{5x}$$

$$= \frac{5}{5x}$$

$$\frac{du}{dx} = \frac{1}{x}$$

$$\frac{dy}{du} = 2u$$

$$\frac{dy}{dx} = \frac{dy}{du} \times \frac{du}{dx}$$

$$= 2u \times \frac{1}{x}$$

$$= \frac{2u}{x}$$

$$\frac{dy}{dx} = \frac{2}{x}\log_e 5x \quad \text{(Since } u = \log_e 5x\text{)}$$

10. If $y = \ln\sqrt{3x^2 - 4}$, find $\frac{dy}{dx}$.

Solution

$$y = \ln\sqrt{3x^2 - 4}$$

Note that $\ln\sqrt{3x^2 - 4}$ is also the same as $\log_e\sqrt{3x^2 - 4}$

Hence, $y = \log_e\sqrt{3x^2 - 4}$

Or, $y = \log_e(3x^2 - 4)^{\frac{1}{2}}$

Let $u = (3x^2 - 4)^{\frac{1}{2}}$

Hence, $y = \log_e u$

$$\frac{du}{dx} = \frac{1}{2}(3x^2 - 4)^{\frac{1}{2} - 1} \times 6x \quad \text{(Note that } 6x \text{ is from the derivative of } 3x^2 - 4\text{)}$$

$$= \frac{6x}{2}(3x^2 - 4)^{-\frac{1}{2}}$$

$$\frac{du}{dx} = \frac{3x}{(3x^2 - 4)^{\frac{1}{2}}}$$

$$\frac{dy}{du} = \frac{1}{u}$$

256

$$\frac{dy}{dx} = \frac{dy}{du} \times \frac{du}{dx}$$

$$= \frac{1}{u} \times \frac{3x}{(3x^2-4)^{\frac{1}{2}}}$$

$$= \frac{1}{(3x^2-4)^{\frac{1}{2}}} \times \frac{3x}{(3x^2-4)^{\frac{1}{2}}}$$

$$\frac{dy}{dx} = \frac{3x}{3x^2-4} \quad \text{(Note that } (3x^2-4)^{\frac{1}{2}} \times (3x^2-4)^{\frac{1}{2}} = 3x^2-4, \text{ after adding their exponents)}$$

11. Find $\frac{dy}{dx}$ if y = $\sqrt[3]{x}$ ln2x.

Solution

y = $\sqrt[3]{x}$ ln2x

We are going to apply product rule here.

$$u = \sqrt[3]{x} = x^{\frac{1}{3}}$$

v = ln2x   (Note that ln2x is the same as $\log_e 2x$)

$$\frac{du}{dx} = \frac{1}{3}x^{\frac{1}{3}-1}$$

$$= \frac{1}{3}x^{-\frac{2}{3}}$$

$$\frac{dv}{dx} = \frac{2}{2x}$$

$$= \frac{1}{x}$$

Hence, $\frac{dy}{dx} = u\frac{dv}{dx} + v\frac{du}{dx}$ (Product rule)

$$= x^{\frac{1}{3}}\left(\frac{1}{x}\right) + \text{ln}2x\left(\frac{1}{3}x^{-\frac{2}{3}}\right)$$

$$= x^{\frac{1}{3}}(x^{-1}) + \frac{1}{3}x^{-\frac{2}{3}}\text{ln}2x$$

$$= x^{-\frac{2}{3}}\left(1 + \frac{\text{ln}2x}{3}\right)$$

$$= x^{-\frac{2}{3}}\left(\frac{3 + \text{ln}2x}{3}\right)$$

$$\frac{dy}{dx} = \frac{1}{x^{\frac{2}{3}}}\left(\frac{3 + \text{ln}2x}{3}\right)$$

Or, $\frac{dy}{dx} = \frac{1}{\sqrt[3]{x^2}}\left(\frac{3 + \text{ln}2x}{3}\right)$

12. Find $\frac{dy}{dx}$ given that y = ln(1 + 2x)^2

Solution

y = ln(1 + 2x)^2 [Also y = $\log_e(1 + 2x)^2$]

257

Let $u = (1 + 2x)^2$

Hence, $y = \ln u$

$$\frac{du}{dx} = 2(1 + 2x) \times \frac{d(2x)}{dx}$$

$$= 2(1 + 2x) \times 2x$$

$$\frac{du}{dx} = 4(1 + 2x)$$

$$\frac{dy}{du} = \frac{1}{u}$$

$$\frac{dy}{dx} = \frac{1}{u} \times 4(1 + 2x)$$

$$= \frac{4(1+2x)}{u}$$

$$= \frac{4(1+2x)}{(1+2x)^2}$$

$$= \frac{4}{1+2x}$$

13. Differentiate with respect to x: $5x\ln(3x^2 - 2)$

Solution

$y = 5x\ln(3x^2 - 2)$

By using product rule:

$u = 5x$

$v = \ln(3x^2 - 2)$

$$\frac{du}{dx} = 5$$

$$\frac{dv}{dx} = \frac{6x}{3x^2 - 2}$$

Hence, $\dfrac{dy}{dx} = u\dfrac{dv}{dx} + v\dfrac{du}{dx}$

$$= 5x\left(\frac{6x}{3x^2-2}\right) + \ln(3x^2 - 2)(5)$$

$$\frac{dy}{dx} = \frac{30x^2}{3x^2-2} + 5\ln(3x^2 - 2)$$

14. Find the derivative of $\dfrac{\ln x}{x}$

Solution

$y = \dfrac{\ln x}{x}$

We are going to use quotient rule as follows:

$u = \ln x$

$v = x$

$$\frac{du}{dx} = \frac{1}{x}$$

$$\frac{dv}{dx} = 1$$

Hence, $\dfrac{dy}{dx} = \dfrac{v\dfrac{du}{dx} - u\dfrac{dv}{dx}}{v^2}$

$$= \frac{x\left(\frac{1}{x}\right) - \ln x \,(1)}{x^2}$$

$$\frac{dy}{dx} = \frac{1 - \ln x}{x^2}$$

Exercise 23

1. Find the derivative of $\log_a 7x$.

2. Find the derivative of $\log_a(x^3 + 5)$

3. Differentiate $\log_a(2x^3 - 5)^6$ with respect to x

4. If $y = \log_a x^{\frac{2}{3}}$, find $\dfrac{dy}{dx}$

5. Find $\dfrac{dy}{dx}$ given that $y = \log_a \dfrac{1 + x^2}{1 - x^2}$

6. If $y = \log_5(5x^3 - 1)$, find $\dfrac{dy}{dx}$.

7. Find $\dfrac{dy}{dx}$ if $y = \log_2 \dfrac{1}{x^2}$

8. Find the derivative of $\log_e(x^2 + 3)$

9. Find the derivative of $(\log_e x)^3$

10. If $y = \ln\sqrt{1 - 2x^5}$, find $\dfrac{dy}{dx}$.

11. Find $\dfrac{dy}{dx}$ if $y = x^4 \ln x^2$

12. Find $\dfrac{dy}{dx}$ given that $y = \ln(3 - 7x)^4$

13. Differentiate with respect to x: $3x^2\ln(4x^3 + 1)$

14. Find the derivative of $\dfrac{\ln x^2}{x^2}$

15. Find the derivative of $\log_e(1 - 5x^2)^3$

CHAPTER 24
DERIVATIVE OF EXPONENTIAL FUNCTIONS

If $y = a^x$, then $\dfrac{dy}{dx} = a^x \log_e a$, where 'a' is any number.

If $y = e^x$, then $\dfrac{dy}{dx} = e^x$

Examples

1. Differentiate with respect to x: a^{5x}

Solution

$\quad y = a^{5x}$

Let $u = 5x$

$\quad y = a^u$

$\dfrac{du}{dx} = 5$

$\dfrac{dy}{du} = a^u \log_e a$

Hence, $\dfrac{dy}{dx} = \dfrac{dy}{du} \quad \times \quad \dfrac{du}{dx}$

$\quad = a^u \log_e a \ \times \ 5$

$\quad = 5a^u \log_e a$

$\dfrac{dy}{dx} = 5a^{5x} \log_e a \qquad$ (since u = 5x)

2. Find the derivative of $a^{x^2 - 3x + 4}$

Solution

$\quad y = a^{x^2 - 3x + 4}$

Let us solve this example directly without using u = $x^2 - 3x + 4$

$\dfrac{dy}{dx} = \dfrac{d(x^2 - 3x + 4)}{dx} \times a^{x^2 - 3x + 4} \times \log_e a$

$\quad = (2x - 3)(a^{x^2 - 3x + 4} \times \log_e a)$

3. If $y = 3x^2 a^{5x}$, find $\dfrac{dy}{dx}$.

Solution

$\quad y = 3x^2 a^{5x}$

We apply product rule as follows:

$\quad u = 3x^2$

$\quad v = a^{5x}$

$$\frac{du}{dx} = 6x$$

$$\frac{dv}{dx} = 5a^{5x}\log_e a$$

Hence, $\dfrac{dy}{dx} = u\dfrac{dv}{dx} + v\dfrac{du}{dx}$

$$= 3x^2(5a^{5x}\log_e a) + a^{5x}(6x)$$

$$= 15x^2 a^{5x}\log_e a + 6xa^{5x}$$

Factorizing the expression above gives:

$$\frac{dy}{dx} = 3xa^{5x}(5x\log_e a + 2)$$

4. Find the derivative of $\dfrac{e^x + e^{-x}}{e^x - e^{-x}}$

Solution

$$y = \frac{e^x + e^{-x}}{e^x - e^{-x}}$$

$$u = e^x + e^{-x}$$

$$v = e^x - e^{-x}$$

$$\frac{du}{dx} = e^x - e^{-x} \quad \text{(The derivative of } e^{-x} = -1 \times e^{-x} = -e^{-x}. \text{ The value } -1 \text{ is from the derivative of } -x)$$

$$\frac{dv}{dx} = e^x + e^{-x}$$

Hence, $\dfrac{dy}{dx} = \dfrac{v\dfrac{du}{dx} - u\dfrac{dv}{dx}}{v^2}$ (Quotient rule)

$$= \frac{(e^x - e^{-x})(e^x - e^{-x}) - (e^x + e^{-x})(e^x + e^{-x})}{(e^x - e^{-x})^2}$$

Expanding the numerator and denominator gives:

$$\frac{dy}{dx} = \frac{(e^x)(e^x) - (e^x)(e^{-x}) - (e^{-x})(e^x) + (e^{-x})(e^{-x}) - [(e^x)(e^x) + (e^x)(e^{-x}) + (e^{-x})(e^x) + (e^{-x})(e^{-x})]}{(e^x - e^{-x})^2}$$

$$= \frac{e^{2x} - 1 - 1 + e^{-2x} - (e^{2x} + 1 + 1 + e^{-2x})}{(e^x - e^{-x})^2} \quad \text{(Note that exponents have been added, and } e^0 = 1)$$

$$= \frac{e^{2x} - 2 + e^{-2x} - e^{2x} - 1 - 1 - e^{-2x})}{(e^x - e^{-x})^2}$$

$$\frac{dy}{dx} = \frac{-4}{(e^x - e^{-x})^2} \quad (e^{2x} \text{ and } e^{-2x} \text{ cancel out})$$

5. If $y = e^{\sqrt{x}}$ find $\dfrac{dy}{dx}$.

Solution

$$y = e^{\sqrt{x}} = e^{x^{\frac{1}{2}}}$$

$$\frac{dy}{dx} = \frac{d(x)^{\frac{1}{2}}}{dx} \times e^{x^{\frac{1}{2}}}$$

$$= \frac{1}{2} x^{-\frac{1}{2}} \times e^{x^{\frac{1}{2}}}$$

$$= \frac{1}{2}\left(\frac{1}{x^{\frac{1}{2}}}\right) \times e^{x^{\frac{1}{2}}}$$

$$= \frac{e^{x^{\frac{1}{2}}}}{2x^{\frac{1}{2}}}$$

$$\frac{dy}{dx} = \frac{e^{\sqrt{x}}}{2\sqrt{x}}$$

6. Find $\frac{dy}{dx}$ if $y = x^{\frac{1}{2}} e^{x^{\frac{1}{2}}}$

Solution

$$y = x^{\frac{1}{2}} e^{x^{\frac{1}{2}}}$$

From product rule:

$$\frac{dy}{dx} = x^{\frac{1}{2}} \frac{d(e^{x^{\frac{1}{2}}})}{dx} + e^{x^{\frac{1}{2}}} \frac{d(x^{\frac{1}{2}})}{dx}$$

$$= x^{\frac{1}{2}} \left(\frac{e^{x^{\frac{1}{2}}}}{2x^{\frac{1}{2}}}\right) + e^{x^{\frac{1}{2}}} \left(\frac{1}{2} x^{-\frac{1}{2}}\right) \quad \text{(Note that the derivative of } e^{x^{\frac{1}{2}}} \text{ is } \frac{e^{x^{\frac{1}{2}}}}{2x^{\frac{1}{2}}} \text{ from example 5)}$$

$$= \frac{e^{x^{\frac{1}{2}}}}{2} + \frac{e^{x^{\frac{1}{2}}}}{2x^{\frac{1}{2}}}$$

$$= \frac{x^{\frac{1}{2}}e^{x^{\frac{1}{2}}} + e^{x^{\frac{1}{2}}}}{2x^{\frac{1}{2}}}$$

$$\frac{dy}{dx} = \frac{e^{x^{\frac{1}{2}}}(x^{\frac{1}{2}} + 1)}{2x^{\frac{1}{2}}}$$

7. Find the derivative of $a^{2x} - a^{-2x}$

Solution

$$y = a^{2x} - a^{-2x}$$

$$\frac{dy}{dx} = \frac{d(2x)}{dx}(a^{2x}\log_e a) - \left[\frac{d(-2x)}{dx}(a^{-2x}\log_e a)\right]$$

$$= 2a^{2x}\log_e a - (-2a^{-2x}\log_e a)$$

$$= 2a^{2x}\log_e a + 2a^{-2x}\log_e a$$

$$= 2\log_e a(a^{2x} + a^{-2x})$$

8. Find the derivative of $e^{2x}\log_e 3x$

Solution

$$y = e^{2x}\log_e 3x$$

By product rule:

$$u = e^{2x}$$

$$v = \log_e 3x \quad (\text{or } v = \ln 3x)$$

$$\frac{du}{dx} = 2e^{2x}$$

$$\frac{dv}{dx} = \frac{3}{3x}$$

$$= \frac{1}{x}$$

$$\frac{dy}{dx} = e^{2x}\left(\frac{1}{x}\right) + \log_e 3x(2e^{2x})$$

$$= \frac{e^{2x}}{x} + 2e^{2x}\log_e 3x$$

$$\frac{dy}{dx} = e^{2x}\left(\frac{1}{x} + 2\log_e 3x\right)$$

9. Differentiate with respect to x: $2\sqrt{xe^x}$

Solution

$$y = 2\sqrt{xe^x}$$

Or, $y = 2\sqrt{x}\sqrt{e^x}$

From product rule:

$$u = 2\sqrt{x} = 2x^{\frac{1}{2}}$$

$$v = \sqrt{e^x} = (e^x)^{\frac{1}{2}}$$

$$v = e^{\frac{1}{2}x} \quad (\text{When the exponents are multiplied})$$

$$\frac{du}{dx} = \frac{1}{2} \times 2x^{-\frac{1}{2}}$$

$$= x^{-\frac{1}{2}}$$

$$\frac{dv}{dx} = \frac{1}{2}e^{\frac{1}{2}x} \quad (\text{Note that } \frac{1}{2} \text{ is from the derivative of } \frac{1}{2}x)$$

$$\frac{dy}{dx} = u\frac{dv}{dx} + v\frac{du}{dx}$$

$$= 2x^{\frac{1}{2}}\left(\frac{1}{2}e^{\frac{1}{2}x}\right) + e^{\frac{1}{2}x}\left(x^{-\frac{1}{2}}\right)$$

$$= \left(\frac{1}{2} \times 2x^{\frac{1}{2}} \times e^{\frac{1}{2}x}\right) \times \frac{e^{\frac{1}{2}x}}{x^{\frac{1}{2}}}$$

$$= x^{\frac{1}{2}} e^{\frac{1}{2}x} + \frac{e^{\frac{1}{2}x}}{x^{\frac{1}{2}}}$$

$$= \frac{x\, e^{\frac{1}{2}x} + e^{\frac{1}{2}x}}{x^{\frac{1}{2}}}$$

$$= \frac{e^{\frac{1}{2}x}(x+1)}{x^{\frac{1}{2}}}$$

$$= \frac{e^{\frac{x}{2}}(x+1)}{x^{\frac{1}{2}}}$$

$$\frac{dy}{dx} = \frac{\sqrt{e^x}\,(x+1)}{\sqrt{x}}$$

10. Find the derivative of $2x\log_{10}x$

<u>Solution</u>

$y = 2x\log_{10}x$ (Note that this is similar to $2x\log_a x$)

$u = 2x$

$v = \log_{10}x$

$\dfrac{du}{dx} = 2$

$\dfrac{dv}{dx} = \dfrac{1}{x}\log_{10}e$ (Recall that the derivative of $\log_a x$ is $\dfrac{1}{x}\log_a e$)

$\dfrac{dy}{dx} = u\dfrac{dv}{dx} + v\dfrac{du}{dx}$

$\quad = 2x\left(\dfrac{1}{x}\log_{10}e\right) + \log_{10}x(2)$

$\quad = 2\log_{10}e + 2\log_{10}x$

$\dfrac{dy}{dx} = 2(\log_{10}e + \log_{10}x)$

Or, $\dfrac{dy}{dx} = 2\left(\dfrac{\log_e e}{\log_e 10} + \log_{10}x\right)$

Note that the rule of change of base in logarithm that has been applied above is given by:

If $\log_a b$ is to be converted to a new base, e, then: $\log_a b = \dfrac{\log_e b}{\log_e a}$

$\dfrac{dy}{dx} = 2\left(\dfrac{1}{\log_e 10} + \log_{10}x\right)$ (Recall that $\log_x x = 1$, or $\log_e e = 1$)

11. Find the derivative of 10^{2x}

<u>Solution</u>

$y = 10^{2x}$ (This is similar to a^{2x})

$\dfrac{dy}{dx} = \dfrac{d(2x)}{dx} \times 10^{2x}\log_e 10$

264

$$= 2 \times 10^{2x} \log_e 10$$

$$= (10^{2x}) 2 \log_e 10$$

$$= 10^{2x} \log_e 10^2 \qquad \text{(Recall that } a \log_x y = \log_x y^a)$$

$$\frac{dy}{dx} = 10^{2x} \log_e 100$$

12. If $y = e^{\ln x}$ find $\frac{dy}{dx}$

<u>Solution</u>

$$y = e^{\ln x} \qquad \text{(Or } y = e^{\log_e x} \text{ since } \ln x = \log_e x)$$

Let, $u = \ln x$

$$y = e^u$$

$$\frac{du}{dx} = \frac{1}{x}$$

$$\frac{dy}{du} = e^u$$

$$\frac{dy}{dx} = e^u \times \frac{1}{x}$$

$$= \frac{e^u}{x}$$

$$= \frac{e^{\ln x}}{x} \qquad \text{(Since } u = \ln x)$$

$$= \frac{e^{\log_e x}}{x} \qquad \text{(Since } \ln x = \log_e x)$$

$$= \frac{x}{x} \qquad \text{(Recall the identity that } m^{\log_m n} = n)$$

$$\frac{dy}{dx} = 1$$

This example shows that calculus works, since the derivative of x is 1, and the derivative of $e^{\log_e x} = 1$ because $e^{\log_e x} = x$. Hence, we could have solved this example by stating that:

$$\frac{d(e^{\ln x})}{dx} = \frac{d(x)}{dx} = 1$$

13. What is the derivative of $\dfrac{e^{\frac{1}{x}}}{x^2}$

<u>Solution</u>

$$y = \frac{e^{\frac{1}{x}}}{x^2}$$

From quotient rule:

$$u = e^{\frac{1}{x}}$$

$$v = x^2$$

$$\frac{du}{dx} = \frac{d\left(\frac{1}{x}\right)}{dx} \times e^{\frac{1}{x}}$$

$$= \frac{d(x^{-1})}{dx} \times e^{\frac{1}{x}}$$

$$= -1x^{-2} \times e^{\frac{1}{x}}$$

$$= \frac{-1}{x^2} \times e^{\frac{1}{x}}$$

$$\frac{du}{dx} = \frac{-e^{\frac{1}{x}}}{x^2}$$

$$\frac{du}{dx} = 2x$$

$$\frac{dy}{dx} = \frac{v\frac{du}{dx} - u\frac{dv}{dx}}{v^2} \qquad \text{(Quotient rule)}$$

$$= \frac{x^2\left(\frac{-e^{\frac{1}{x}}}{x^2}\right) - e^{\frac{1}{x}}(2x)}{(x^2)^2}$$

$$= \frac{-e^{\frac{1}{x}} - 2x\, e^{\frac{1}{x}}}{x^4} \qquad \text{(Note that } x^2 \text{ has cancelled out)}$$

14. Find the derivative of $e^{-5x} + 5e$

<u>Solution</u>

$$y = e^{-5x} + 5e$$

$$\frac{dy}{dx} = \frac{d(-5x)}{dx} \times e^{-5x} + \frac{d(5e)}{dx}$$

$$= -5e^{-5x} + 0 \qquad \text{(Note that 5e is a constant and the derivative of a constant is zero)}$$

$$\frac{dy}{dx} = -5e^{-5x}$$

15. If $y = e^{(2x+3)^2}$ find $\dfrac{dy}{dx}$.

<u>Solution</u>

$$y = e^{(2x+3)^2}$$

Let $u = (2x + 3)^2$

Hence, $y = e^{u}$

$$\frac{du}{dx} = 2(2x + 3)^{2-1} \times \frac{d(2x)}{dx} \qquad \text{(Using chain rule)}$$

$$= 2(2x + 3) \times 2$$

$$= 4(2x + 3)$$

$$\frac{dy}{du} = e^{u}$$

$$\frac{dy}{dx} = \frac{dy}{du} \times \frac{du}{dx}$$

$$= e^u \times 4(2x + 3)$$

$$= e^{(2x+3)^2} \times 4(2x + 3) \qquad \text{(Note that } u = (2x + 3)^2\text{)}$$

$$\frac{dy}{dx} = 4(2x + 3)e^{(2x+3)^2}$$

16. Differentiate with respect to θ, $5^{2\theta}$

Solution

$$y = 5^{2\theta} \qquad \text{(This is like } y = a^{2\theta}\text{)}$$

$$\frac{dy}{d\theta} = \frac{d(2\theta)}{d\theta} \times 5^{2\theta} \log_e 5$$

$$= 2 \times 5^{2\theta} \log_e 5$$

$$= (5^{2\theta})2\log_e 5$$

$$= 5^{2\theta} \log_e 5^2 \qquad \text{(Note that } 2\log_e 5 = \log_e 5^2\text{)}$$

$$\frac{dy}{d\theta} = 5^{2\theta} \log_e 25$$

Exercise 24

1. Differentiate with respect to x: $5a^{2x}$

2. Find the derivative of $a^{5x^3 - x}$

3. If $y = x^4 a^{3x}$, find $\frac{dy}{dx}$.

4. Find the derivative of $\dfrac{3 + e^{-x}}{e^x}$

5. If $y = 2e^{\sqrt{3x}}$ find $\frac{dy}{dx}$.

6. Find $\frac{dy}{dx}$ if $y = 6x^3 e^{x^2}$

7. Find the derivative of $a^{-5x} + a^{5x}$

8. Find the derivative of $3e^{4x}\log_e 2x^2$

9. Differentiate with respect to x: $3x^2\sqrt{e^{5x}}$

10. Find the derivative of $x^4 \log_{10} 7x$

11. Find the derivative of 6^{3x}

12. Find the derivative of $e^{x^2 - 3x}$

13. What is the derivative of $\dfrac{2e^{\frac{3}{x}}}{10x^5}$

14. Find the derivative of $1 - 2e^{-3x}$

15. If $y = e^{(x^2+x)^3}$ find $\dfrac{dy}{dx}$.

16. Differentiate with respect to x, 2^{x^2}

17. Find the derivative of $\dfrac{a^x - a^{-x}}{a^{-x}}$

18. Find the derivative of $e^x \ln 10x^3$

19. Differentiate with respect to x: $e^{2x}\sqrt{5x}$

20. Find the derivative of $2x^5 \log_{10} 5x^4$

CHAPTER 25
LOGARITHMIC DIFFERENTIATION

A function of x could be raised to an exponent which is also a function of x. An example is x^x. The differentiation of such a function is called logarithmic differentiation.

This type of function is differentiated by first taking the logarithm of both sides of the equation, and then differentiating implicitly before making $\frac{dy}{dx}$ the subject of the formula (i.e. solving for $\frac{dy}{dx}$). Finally, y (i.e. the function) is substituted into the final answer.

Examples

1. Find the derivative of x^x.

<u>Solution</u>

$$y = x^x$$

This is a function of x raised to an exponent which is also a function of x.

Hence, take the logarithm to base e of both sides as follows:

$$\log_e y = \log_e x^x$$
$$\log_e y = x\log_e x \quad \text{(Note that } \log_m n^b = b\log_m n\text{)}$$

Now differentiate y implicitly and differentiate the right hand side using product rule. This gives:

$$\frac{1}{y}\frac{dy}{dx} = x\frac{1}{x} + \log_e x(1) \quad \text{(Use of product rule on the right hand side, with u = } x \text{ and v = } \log_e x\text{)}$$

$$\frac{1}{y}\frac{dy}{dx} = 1 + \log_e x$$

Multiply both sides by y to make $\frac{dy}{dx}$ the subject of the formula as follows:

$$\frac{dy}{dx} = y(1 + \log_e x)$$
$$\frac{dy}{dx} = x^x(1 + \log_e x) \quad \text{(Since y = } x^x\text{)}$$

2. Find the derivative of $x^{\ln x}$.

<u>Solution</u>

$$y = x^{\ln x}$$

Taking the logarithm of both sides gives:

$$\log_e y = \log_e x^{\ln x}$$
$$\log_e y = \ln x\log_e x$$

Differentiate y implicitly and treat the right hand side using product rule as follows:

$$\frac{1}{y}\frac{dy}{dx} = \ln x\left(\frac{1}{x}\right) + \log_e x\left(\frac{1}{x}\right)$$

$$\frac{1}{y}\frac{dy}{dx} = \left(\frac{\ln x}{x}\right) + \left(\frac{\ln x}{x}\right) \qquad \text{(Note that } \log_e x \text{ can also be written as } \ln x)$$

$$\frac{1}{y}\frac{dy}{dx} = \frac{2\ln x}{x}$$

Multiply both sides by y. This gives:

$$\frac{dy}{dx} = y\left(\frac{2\ln x}{x}\right)$$

$$\frac{dy}{dx} = x^{\ln x}\left(\frac{2\ln x}{x}\right) \qquad \text{(Since } y = x^{\ln x})$$

Or, $\dfrac{dy}{dx} = \dfrac{2}{x} x^{\ln x} \ln x$

3. If $y = \ln(2+x)^x$ find $\dfrac{dy}{dx}$.

Solution

$$y = \ln(2+x)^x$$

In this case we are not going to take the logarithm of both sides since there is already logarithm on the right hand side. Hence, we proceed as follows:

$$y = \ln(2+x)^x$$
$$y = x\log_e(2+x) \qquad \text{[Note that } \ln(2+x) \text{ can also be written as } \log_e(2+x)]$$

Using product rule gives:

$$\frac{dy}{dx} = x\left(\frac{1}{2+x}\right) + \log_e(2+x)(1)$$

$$\frac{dy}{dx} = \frac{x}{2+x} + \log_e(2+x)$$

4. Find $\dfrac{dy}{dx}$ if $y = (x^2 - 3)^{5x}$

Solution

$$y = (x^2 - 3)^{5x}$$

Taking the natural logarithm of both sides gives:

$$\log_e y = \log_e(x^2 - 3)^{5x}$$
$$\log_e y = 5x\log_e(x^2 - 3)$$

Using implicit differentiation for the left hand side and product rule for the right hand side gives:

$$\frac{1}{y}\frac{dy}{dx} = 5x\left(\frac{2x}{x^2-3}\right) + \log_e(x^2 - 3)(5)$$

$$\frac{1}{y}\frac{dy}{dx} = \frac{10x^2}{x^2-3} + 5\log_e(x^2 - 3)$$

Multiply both sides by y to obtain:

$$\frac{dy}{dx} = y\left[\frac{10x^2}{x^2-3} + 5\log_e(x^2 - 3)\right]$$

$$\frac{dy}{dx} = (x^2 - 3)^{5x} \left[\frac{10x^2}{x^2 - 3} + 5\log_e(x^2 - 3) \right] \qquad \text{[Note that } y = (x^2 - 3)^{5x}\text{]}$$

5. If $y = \dfrac{(x^2 - 5)(3x - 1)^2}{x^7(2x^3 - 3)}$

<u>Solution</u>

This is not a function of x raised to an exponent which is also a function of x. This function can be differentiated by the use of product and quotient rule. However, the use of these rules will be a nightmare. Hence, we have to apply logarithmic differentiation. This is done as follows:

$$y = \frac{(x^2 - 5)(3x - 1)^2}{x^7(2x^3 - 3)}$$

Taking the logarithm of both sides gives:

$$\log_e y = \log_e \left[\frac{(x^2 - 5)(3x - 1)^2}{x^7(2x^3 - 3)} \right]$$

In order to continue, we have to recall the theory of logarithm as follows:

1. $\log_b(cd) = \log_b c + \log_b d$

2. $\log_b\left(\frac{c}{d}\right) = \log_b c - \log_b d$

3. $\log_b c^d = d\log_b c$

Therefore, we continue the logarithmic differentiation as follows:

$$\log_e y = \log_e \left[\frac{(x^2 - 5)(3x - 1)^2}{x^7(2x^3 - 3)} \right]$$

Applying the theory above gives:

$$\log_e y = \log_e[(x^2 - 5)(3x - 1)^2] - \log_e[x^7(2x^3 - 3)]$$
$$\log_e y = \log_e(x^2 - 5) + \log_e(3x - 1)^2 - [\log_e x^7 + \log_e(2x^3 - 3)]$$
$$\log_e y = \log_e(x^2 - 5) + 2\log_e(3x - 1) - 7\log_e x - \log_e(2x^3 - 3)$$

Differentiating each term accordingly gives:

$$\frac{1}{y}\frac{dy}{dx} = \frac{2x}{x^2 - 5} + \frac{2 \times 3}{3x - 1} - \frac{7}{x} - \frac{6x}{2x^3 - 3}$$

Note that each function of x is differentiated first, and the value obtained is then divided by the original function. Hence, the expression above simplifies to gives:

$$\frac{1}{y}\frac{dy}{dx} = \frac{2x}{x^2 - 5} + \frac{6}{3x - 1} - \frac{7}{x} - \frac{6x}{2x^3 - 3}$$

Multiply both sides by y to obtain $\dfrac{dy}{dx}$ as follows:

$$\frac{dy}{dx} = y\left(\frac{2x}{x^2 - 5} + \frac{6}{3x - 1} - \frac{7}{x} - \frac{6x}{2x^3 - 3} \right)$$

Now replace y with its original value from the question as follows:

$$\frac{dy}{dx} = \left[\frac{(x^2 - 5)(3x - 1)^2}{x^7(2x^3 - 3)} \right]\left(\frac{2x}{x^2 - 5} + \frac{6}{3x - 1} - \frac{7}{x} - \frac{6x}{2x^3 - 3} \right)$$

6. If $y = \dfrac{(2x - 1)(x + 4)(6x - 5)}{(3 - x)^2}$

Solution

$y = \dfrac{(2x - 1)(x + 4)(6x - 5)}{(3 - x)^2}$

This is a complex function where product rule and quotient rule would be difficult to apply.

Hence, we apply logarithmic differentiation as follows:

$\log_e y = \log_e \left[\dfrac{(2x - 1)(x + 4)(6x - 5)}{(3 - x)^2} \right]$

Applying the theory of logarithm gives:

$\log_e y = \log_e[(2x - 1)(x + 4)(6x - 5)] - \log_e(3 - x)^2$

$\log_e y = \log_e(2x - 1) + \log_e(x + 4) + \log_e(6x - 5) - 2\log_e(3 - x)$

Differentiating accordingly gives:

$\dfrac{1}{y}\dfrac{dy}{dx} = \dfrac{2}{2x - 1} + \dfrac{1}{x + 4} + \dfrac{6}{6x - 5} - \dfrac{2(-1)}{3 - x}$

$\dfrac{1}{y}\dfrac{dy}{dx} = \dfrac{2}{2x - 1} + \dfrac{1}{x + 4} + \dfrac{6}{6x - 5} + \dfrac{2}{3 - x}$

Multiply both sides by y. This gives:

$\dfrac{dy}{dx} = y\left(\dfrac{2}{2x - 1} + \dfrac{1}{x + 4} + \dfrac{6}{6x - 5} + \dfrac{2}{3 - x} \right)$

Finally replace y with its original expression as follows:

$\dfrac{dy}{dx} = \left[\dfrac{(2x - 1)(x + 4)(6x - 5)}{(3 - x)^2} \right] \left(\dfrac{2}{2x - 1} + \dfrac{1}{x + 4} + \dfrac{6}{6x - 5} + \dfrac{2}{3 - x} \right)$

7. Find the derivative of x^{x^2}

Solution

$y = x^{x^2}$

$\log_e y = \log_e x^{x^2}$

$\log_e y = x^2 \log_e x$

$\dfrac{1}{y}\dfrac{dy}{dx} = x^2\left(\dfrac{1}{x}\right) + \log_e x(2x)$

$\dfrac{1}{y}\dfrac{dy}{dx} = x + 2x\log_e x$

$\dfrac{1}{y}\dfrac{dy}{dx} = x(1 + 2\log_e x)$

$\dfrac{dy}{dx} = yx(1 + 2\log_e x)$

$= x^{x^2} x(1 + 2\log_e x)$ (y has been replaced with x^{x^2})

$\dfrac{dy}{dx} = x^{x^2 + 1}(1 + 2\log_e x)$

Note that the exponents of x^{x^2} and x were added to obtain $x^{x^2 + 1}$ since x is also x^1

8. If $y = e^{e^x}$ find $\dfrac{dy}{dx}$.

Solution

$y = e^{e^x}$

$\log_e y = \log_e e^{e^x}$

$\log_e y = e^x \log_e e$

$\log_e y = e^x$ (Recall that logarithm of a number to the same base is 1. Hence, $\log_e e = 1$)

$\dfrac{1}{y}\dfrac{dy}{dx} = e^x$ (Note that the derivative of e^x is e^x)

$\dfrac{dy}{dx} = y e^x$

$\phantom{\dfrac{dy}{dx}} = e^{e^x} e^x$

$\dfrac{dy}{dx} = e^{e^x + x}$ (Their exponents have been added)

9. Find the derivative of 5^{2^x}

Solution

$y = 5^{2^x}$

$\log_e y = \log_e 5^{2^x}$

$\log_e y = 2^x \log_e 5$

Differentiate the left hand side implicitly. On the right hand side, differentiate 2^x and take $\log_e 5$ as a constant. This gives:

$\dfrac{1}{y}\dfrac{dy}{dx} = 2^x \log_e 2 \log_e 5$

Note that 2^x is treated in a similar way as a^x, and recall that the derivative of a^x is $a^x \log_e a$. Hence the derivative of 2^x is $2^x \log_e 2$, and $\log_e 5$ multiplies it since it is a constant. Therefore:

$\dfrac{dy}{dx} = y(2^x \log_e 2 \log_e 5)$

$\phantom{\dfrac{dy}{dx}} = 5^{2^x}(2^x \log_e 2 \log_e 5)$

$\dfrac{dy}{dx} = 5^{2^x}(2^x \log_e 2)(\log_e 5)$

10. If $y = x^{\ln(2x^2 - 5)}$, find $\dfrac{dy}{dx}$.

Solution

$y = x^{\ln(2x^2 - 5)}$

$\log_e y = \log_e x^{\ln(2x^2 - 5)}$

$\log_e y = \ln(2x^2 - 5)\log_e x$

$\dfrac{1}{y}\dfrac{dy}{dx} = \ln(2x^2 - 5)\left(\dfrac{1}{x}\right) + \log_e x\left(\dfrac{4x}{2x^2 - 5}\right)$ (Use of product rule)

273

$$\frac{1}{y}\frac{dy}{dx} = \frac{1}{x}\ln(2x^2 - 5) + \left(\frac{4x}{2x^2 - 5}\right)\log_e x$$

$$\frac{dy}{dx} = y\left[\frac{1}{x}\ln(2x^2 - 5) + \left(\frac{4x}{2x^2 - 5}\right)\log_e x\right]$$

$$\frac{dy}{dx} = x^{\ln(2x^2 - 5)}\left[\frac{\ln(2x^2 - 5)}{x} + \left(\frac{4x}{2x^2 - 5}\right)\log_e x\right]$$

Exercise 25

1. Find the derivative of x^{2x}

2. Find the derivative of $3x^{\ln 2x}$.

3. If $y = \ln(1 - 4x^2)^x$ find $\dfrac{dy}{dx}$.

4. Find $\dfrac{dy}{dx}$ if $y = (6x^2 - 5)^{3x}$

5. If $y = \dfrac{(2x^3 + 1)(x - 2)^3}{3x^2(x^3 - 1)^2}$

6. If $y = \dfrac{(2x - 1)(x^2 - 2)}{(1 - x)(x - 3)^2}$

7. Find the derivative of $5x^{3x^4}$

8. If $y = e^{e^{3x}}$ find $\dfrac{dy}{dx}$.

9. Find the derivative of 2^{e^x}

10. If $y = x^{\ln(x^2 - 4x)}$, find $\dfrac{dy}{dx}$.

11. If $y = \ln(10 + 2x^2)^x$ find $\dfrac{dy}{dx}$.

12. Find $\dfrac{dy}{dx}$ if $y = (3x^3 - 8)^{2x}$

13. If $y = \dfrac{(x + 3)^2(x - 3)^2}{(x^3 - 1)}$

14. If $y = 5x^{(\ln x)^2}$, find $\dfrac{dy}{dx}$.

15. Find the derivative of $10x^{3x^2}$

CHAPTER 26
DERIVATIVE OF ONE FUNCTION WITH RESPECT TO ANOTHER

We can differentiate one function with respect to another as illustrated by the examples shown below.

Examples

1. Differentiate x^{12} with respect to x^7.

Solution

Let $u = x^{12}$

And $v = x^7$

$$\frac{du}{dx} = 12x^{11}$$

$$\frac{dv}{dx} = 7x^6$$

Differentiating x^{12} with respect to x^7 means differentiating u (i.e. x^{12}) with respect to v (i.e. x^7).

This means $\frac{du}{dv}$. This follows the parametric equation rule given by:

$$\frac{du}{dv} = \frac{\frac{du}{dx}}{\frac{dv}{dx}}$$

$$= \frac{12x^{11}}{7x^6}$$

$$= \frac{12}{7}x^5$$

2. Differentiate $2e^x$ with respect to $\ln 2x$.

Solution

The differentiation of $2e^x$ with respect to $\ln 2x$ simply means:

$$\frac{\text{The derivative of } 2e^x}{\text{The derivative of } \ln 2x}$$

Hence, $\frac{d(2e^x)}{dx} = 2e^x$

And, $\frac{d(\ln 2x)}{dx} = \frac{2}{2x} = \frac{1}{x}$

Therefore, the differentiation of $2e^x$ with respect to $\ln 2x$ is given by:

$$\frac{2e^x}{\frac{1}{x}}$$

$$= 2xe^x$$

3. Differntiate $\sin 5x$ with respect to $\cos x$.

Solution

Hence, $\dfrac{d(\sin 5x)}{dx} = 5\cos 5x$

And, $\dfrac{d(\cos x)}{dx} = -\sin x$

Therefore, the differentiation of $\sin 5x$ with respect to $\cos x$ is given by:

$\dfrac{\text{The derivative of } \sin 5x}{\text{The derivative of } \cos x}$

$= \dfrac{5\cos 5x}{-\sin x}$

$= -\dfrac{5\cos 5x}{\sin x}$

4. Find the derivative of $2x^2 - 5$ with respect to $4x - 1$

Solution

This is obtained as follows:

$\dfrac{\dfrac{d(2x^2 - 5)}{dx}}{\dfrac{d(4x - 1)}{dx}}$

$= \dfrac{4x}{4}$

$= x$

Exercise 26

1. Differentiate x^3 with respect to x^5.
2. Differentiate e^{5x} with respect to $\ln x^2$
3. Differentiate $\cos x^2$ with respect to $\sin x$.
4. Find the derivative of $4x^3 - 5x^2$ with respect to $(x - 1)^2$
5. Differentiate $7x^4$ with respect to $3x^2$.
6. Differentiate $2a^x$ with respect to e^x.
7. Differentiate $\tan^2 x$ with respect to $\cos 5x$.
8. Find the derivative of $\ln(x - 1)$ with respect to e^{x-1}
9. Differentiate $\log_e x^5$ with respect to e^{5x}.
10. Differentiate $\ln(e^x - e^{-x})$ with respect to $\ln x^x$

CHAPTER 27
HIGHER DERIVATIVES (SUCCESSIVE DIFFERENTIATION)

If f(x) is differentiated it gives the first derivative denoted by f'(x) or $\frac{dy}{dx}$. If we differentiate f'(x), it gives the second derivative denoted by f''(x) or $\frac{d^2y}{dx^2}$ (read as, dee two y dee x squared). Other higher derivatives such as $\frac{d^3y}{dx^3}, \frac{d^4y}{dx^4}$ etc can also be obtained depending on the function.

Examples

1. Find the first, second and third derivatives of $2x^5 - 3x^4 + 5x^2 - 6$

Solution

$$y = 2x^5 - 3x^4 + 5x^2 - 6$$

$$\frac{dy}{dx} = 10x^4 - 12x^3 + 10x$$

$\frac{d^2y}{dx^2}$ is obtained by differentiating $\frac{dy}{dx}$, which means to differentiate $10x^4 - 12x^3 + 10x$.

Hence, $\frac{d^2y}{dx^2} = 40x^3 - 36x^2 + 10$

$\frac{d^3y}{dx^3}$ is obtained by differentiating $40x^3 - 36x^2 + 10$ as follows:

$$\frac{d^3y}{dx^3} = 120x^2 - 72x$$

In summary, the first derivative $\left(\frac{dy}{dx}\right)$ is $10x^4 - 12x^3 + 10x$, the second derivative $\left(\frac{d^2y}{dx^2}\right)$ is $40x^3 - 36x^2 + 10$, while the third derivative $\left(\frac{d^3y}{dx^3}\right)$ is $120x^2 - 72x$.

2. If $y = \ln x^2$, find $\frac{d^2y}{dx^2}$.

Solution

$$y = \ln x^2$$

$$\frac{dy}{dx} = \frac{2x}{x^2}$$

$$= \frac{2}{x}$$

$$\frac{d^2y}{dx^2} = \frac{d\left(\frac{2}{x}\right)}{dx}$$

$$= \frac{d(2x^{-1})}{dx}$$

$$= -2x^{-2}$$

$$\frac{d^2y}{dx^2} = \frac{-2}{x^2}$$

3. Find $\frac{d^3y}{dx^3}$ given that $y = e^{x^3}$

Solution

$$y = e^{x^3}$$

$$\frac{dy}{dx} = 3x^2e^{x^3} \qquad \text{(Note that } 3x^2 \text{ is from the derivative of } x^3\text{)}$$

We now use product rule to obtain $\frac{d^2y}{dx^2}$ as follows:

$$\frac{d^2y}{dx^2} = 3x^2\left[\frac{d(e^{x^3})}{dx}\right] + e^{x^3}\left[\frac{d(3x^2)}{dx}\right]$$

$$= 3x^2(3x^2e^{x^3}) + e^{x^3}(6x)$$

$$= 9x^4e^{x^3} + 6xe^{x^3}$$

$$\frac{d^2y}{dx^2} = e^{x^3}(9x^4 + 6x)$$

Finally, let us also use product rule to obtain $\frac{d^3y}{dx^3}$ as follows:

$$\frac{d^3y}{dx^3} = e^{x^3}\left[\frac{d(9x^4 + 6x)}{dx}\right] + (9x^4 + 6x)\left[\frac{d(e^{x^3})}{dx}\right]$$

$$= e^{x^3}(36x^3 + 6) + (9x^4 + 6x)(3x^2e^{x^3})$$

$$= 36x^3e^{x^3} + 6e^{x^3} + 27x^6e^{x^3} + 18x^3e^{x^3}$$

$$= 27x^6e^{x^3} + 36x^3e^{x^3} + 18x^3e^{x^3} + 6e^{x^3}$$

$$= 27x^6e^{x^3} + 54x^3e^{x^3} + 6e^{x^3}$$

$$\frac{d^3y}{dx^3} = 3e^{x^3}(9x^6 + 18x^3 + 2)$$

4. If $y = \sin 2x^3$, find the third derivative of y.

Solution

$$y = \sin 2x^3$$

$$\frac{dy}{dx} = \cos 2x^3 \frac{d(2x^3)}{dx} \qquad \text{(Recall that the derivative of } \sin x \text{ is } \cos x\text{)}$$

$$= \cos 2x^3 (6x^2)$$

$$\frac{dy}{dx} = 6x^2\cos 2x^3$$

We now use product rule to obtain $\frac{d^2y}{dx^2}$ as follows:

$$\frac{d^2y}{dx^2} = 6x^2\left[\frac{d(\cos 2x^3)}{dx}\right] + \cos 2x^3\left[\frac{d(6x^2)}{dx}\right]$$

$$= 6x^2\left[-\sin 2x^3 \frac{d(2x^3)}{dx}\right] + \cos 2x^3(12x)$$

$$= 6x^2\left[-\sin 2x^3\,(6x^2)\right] + 12x\cos 2x^3$$

$$\frac{d^2y}{dx^2} = -36x^4\sin 2x^3 + 12x\cos 2x^3$$

Again, the use of product rule gives us $\frac{d^3y}{dx^3}$ as follows:

$$\frac{d^3y}{dx^3} = -36x^4\left[\frac{d\left(\sin 2x^3\right)}{dx}\right] + \sin 2x^3\left[\frac{d\left(-36x^4\right)}{dx}\right] + 12x\left[\frac{d\left(\cos 2x^3\right)}{dx}\right] + \cos 2x^3\left[\frac{d(12x)}{dx}\right]$$

Note that the derivative of $\sin 2x^3$ is $6x^2\cos 2x^3$ as obtained from $\frac{dy}{dx}$. Similarly, the differentiation of $\cos 2x^3$ will give us $-6x^2\sin 2x^3$ (since the derivative of $\sin x$ and $\cos x$ differs only by sign and the interchanging of sin with cos or cos with sin).

We now replace the derivative of $\sin 2x^3$ and $\cos 2x^3$ with $6x^2\cos 2x^3$ and $-6x^2\sin 2x^3$ respectively in the expression above. This gives:

$$\frac{d^3y}{dx^3} = -36x^4\,(6x^2\cos 2x^3) + \sin 2x^3(-144x^3) + 12x\,(-6x^2\sin 2x^3) + \cos 2x^3(12)$$

$$= -216x^6\cos 2x^3 - 144x^3\sin 2x^3 - 72x^3\sin 2x^3 + 12\cos 2x^3$$

$$\frac{d^3y}{dx^3} = -216x^6\cos 2x^3 - 216x^3\sin 2x^3 + 12\cos 2x^3$$

Or, $\frac{d^3y}{dx^3} = -12(18x^6\cos 2x^3 + 18x^3\sin 2x^3 - \cos 2x^3)$

5. Find $\frac{d^2y}{dx^2}$ given that $y = a^{x^2} - 5$

Solution

$$y = a^{x^2} - 5$$

$$\frac{dy}{dx} = a^{x^2}\left[\frac{d(x^2)}{dx}\right]\log_e a$$

$$= a^{x^2}(2x)\log_e a$$

$$\frac{dy}{dx} = 2x\,a^{x^2}\log_e a$$

We now use product rule to find $\frac{d^2y}{dx^2}$ as follows:

$$\frac{d^2y}{dx^2} = 2x\left[\frac{d\left(a^{x^2}\log_e a\right)}{dx}\right] + a^{x^2}\log_e a\left[\frac{d(2x)}{dx}\right]$$

$$= 2x\left[\frac{\log_e a\, d\left(a^{x^2}\right)}{dx}\right] + a^{x^2}\log_e a(2)$$

$$= 2x[\log_e a(2xa^{x^2}\log_e a)] + 2a^{x^2}\log_e a$$

$$= 4x^2a^{x^2}(\log_e a)^2 + 2a^{x^2}\log_e a$$

$$\frac{d^2y}{dx^2} = 2a^{x^2}\log_e a(2x^2\log_e a \; + \; 1)$$

6. Find the second derivative of $y = \dfrac{\cos x}{x^3}$

Solution

$$y = \frac{\cos x}{x^3}$$

We use quotient rule as follows:

$$\frac{dy}{dx} = \frac{x^3\frac{d(\cos x)}{dx} - \cos x\frac{d(x^3)}{dx}}{(x^3)^2}$$

$$= \frac{x^3(-\sin x) - \cos x(3x^2)}{x^6}$$

$$\frac{dy}{dx} = \frac{-x^3\sin x - 3x^2\cos x}{x^6}$$

Divide each part by x^6 to separate into fractions as follows:

$$\frac{dy}{dx} = \frac{-x^3\sin x}{x^6} - \frac{3x^2\cos x}{x^6}$$

$$\frac{dy}{dx} = \frac{-\sin x}{x^3} - \frac{3\cos x}{x^4}$$

We now apply quotient rule again as follows:

$$\frac{d^2y}{dx^2} = \frac{x^3\frac{d(-\sin x)}{dx} - (-\sin x)\frac{d(x^3)}{dx}}{(x^3)^2} - \left[\frac{x^4\frac{d(3\cos x)}{dx} - 3\cos x\frac{d(x^4)}{dx}}{(x^4)^2}\right]$$

$$= \frac{x^3(-\cos x) + \sin x\,(3x^2)}{x^6} - \left[\frac{x^4(-3\sin x) - 3\cos x\,(4x^3)}{x^8}\right]$$

$$= \frac{-x^3\cos x + 3x^2\sin x}{x^6} + \frac{3x^4\sin x + 12x^3\cos x}{x^8}$$

Separating into fractions gives:

$$\frac{d^2y}{dx^2} = \frac{-x^3\cos x}{x^6} + \frac{3x^2\sin x}{x^6} + \frac{3x^4\sin x}{x^8} + \frac{12x^3\cos x}{x^8}$$

$$= \frac{-\cos x}{x^3} + \frac{3\sin x}{x^4} + \frac{3\sin x}{x^4} + \frac{12\cos x}{x^5}$$

Combining the fractions again by using x^5 as LCM gives:

$$\frac{d^2y}{dx^2} = \frac{-x^2\cos x + 3x\sin x + 3x\sin x + 12\cos x}{x^5}$$

$$\frac{d^2y}{dx^2} = \frac{6x\sin x + 12\cos x - x^2\cos x}{x^5}$$

7. Find the third derivative of $y = \log_e(1 + 2x)^2$

Solution

$$y = \log_e(1 + 2x)^2$$

$$\frac{dy}{dx} = \frac{2(1+2x)^{2-1} \times \frac{d(2x)}{dx}}{(1+2x)^2}$$ [Note the use of chain rule in finding the derivative of $(1 + 2x)^2$]

$$= \frac{2(1+2x) \times 2}{(1+2x)^2}$$

$$= \frac{4(1+2x)}{(1+2x)^2}$$

$$= \frac{4}{1+2x} \quad (1 + 2x \text{ cancels out})$$

$$\frac{dy}{dx} = 4(1 + 2x)^{-1}$$

Using chain rule gives $\frac{d^2y}{dx^2}$ as follows:

$$\frac{d^2y}{dx^2} = -1 \times 4(1 + 2x)^{-1-1} \times \frac{d(2x)}{dx}$$

$$= -4(1 + 2x)^{-2} \times 2$$

$$= -8(1 + 2x)^{-2}$$

Use of chain rule again gives $\frac{d^3y}{dx^3}$ as follows:

$$\frac{d^3y}{dx^3} = -2 \times -8(1 + 2x)^{-2-1} \times \frac{d(2x)}{dx}$$

$$= 16(1 + 2x)^{-3} \times 2$$

$$= 32(1 + 2x)^{-3}$$

$$\frac{d^3y}{dx^3} = \frac{32}{(1+2x)^3}$$

8. If y = sec2x, find $\frac{d^3y}{dx^3}$.

<u>Solution</u>

    y = sec2x

$$\frac{dy}{dx} = 2\sec2x\tan2x$$

We now use product rule to obtain $\frac{d^2y}{dx^2}$ as follows:

$$\frac{d^2y}{dx^2} = 2\sec2x \frac{d(\tan 2x)}{dx} + \tan2x \frac{d(2\sec 2x)}{dx}$$

$$= 2\sec2x(2\sec^2 2x) + \tan2x(2 \times 2 \sec2x\tan2x)$$

$$\frac{d^2y}{dx^2} = 4\sec^3 2x + 4\sec2x\tan^2 2x$$

We now use chain rule for $4\sec^3 2x$ and product rule for $4\sec2x\tan^2 2x$ to obtain $\frac{d^3y}{dx^3}$ as follows:

$$\frac{d^3y}{dx^3} = 3 \times 4 \sec^2 2x \frac{d(\sec 2x)}{dx} + 4\sec2x \frac{d(\tan^2 2x)}{dx} + \tan^2 2x \frac{d(4\sec 2x)}{dx}$$

$$= 12\sec^2 2x(2\sec2x\tan2x) + 4\sec2x \left[2\tan2x \frac{d(\tan2x)}{dx}\right] + \tan^2 2x[4(2\sec2x\tan2x)]$$

281

$$= 24\sec^3 2x\tan 2x + 8\sec 2x\tan 2x(2\sec^2 2x) + 8\sec 2x\tan^3 2x$$
$$= 24\sec^3 2x\tan 2x + 16\sec^3 2x\tan 2x + 8\sec 2x\tan^3 2x$$
$$\frac{d^3y}{dx^3} = 40\sec^3 2x\tan 2x + 8\sec 2x\tan^3 2x$$

9. Given that y = 2cos5x + 3sin5x, prove that $\frac{d^2y}{dx^2}$ + 25y = 0

<u>Solution</u>

y = 2cos5x + 3sin5x

$$\frac{dy}{dx} = 2 \times 5(-\sin 5x) + 3 \times 5(\cos 5x)$$
$$= -10\sin 5x + 15\cos 5x$$
$$\frac{d^2y}{dx^2} = -10 \times 5(\cos 5x) + 15 \times 5(-\sin 5x)$$
$$= -50\cos 5x - 75\sin 5x$$

We now obtain 25y as follows:

y = 2cos5x + 3sin5x

25y = 25(2cos5x + 3sin5x)

25y = 50cos5x + 75sin5x

We now simplify $\frac{d^2y}{dx^2}$ + 25y as follows:

$$\frac{d^2y}{dx^2} + 25y = -50\cos 5x - 75\sin 5x + 50\cos 5x + 75\sin 5x$$
$$= 0 \qquad \text{(Equal terms with opposite signs cancel out)}$$

Therefore, $\frac{d^2y}{dx^2}$ + 25y = 0   (As proven above)

10. If $y = x + \sqrt{4 + x^2}$, show that $(4 + x^2)\frac{d^2y}{dx^2} + x\frac{dy}{dx} - y = 0$

<u>Solution</u>

$$y = x + \sqrt{4 + x^2}$$
$$y = x + (4 + x^2)^{\frac{1}{2}}$$
$$\frac{dy}{dx} = 1 + \frac{1}{2}(4 + x^2)^{\frac{1}{2} - 1} \times \frac{d(x^2)}{dx}$$
$$= 1 + \frac{1}{2}(4 + x^2)^{-\frac{1}{2}} \times 2x$$
$$\frac{dy}{dx} = 1 + x(4 + x^2)^{-\frac{1}{2}}$$

Using product rule, we obtain $\frac{d^2y}{dx^2}$ as follows:

$$\frac{d^2y}{dx^2} = 0 + x\left[\frac{d(4+x^2)^{-\frac{1}{2}}}{dx}\right] + (4+x^2)^{-\frac{1}{2}}\frac{d(x)}{dx}$$

$$= x\left[-\frac{1}{2}(4+x^2)^{-\frac{1}{2}-1}\frac{d(x^2)}{dx}\right] + (4+x^2)^{-\frac{1}{2}}(1)$$

$$= x\left[-\frac{1}{2}(4+x^2)^{-\frac{3}{2}} \times 2x\right] + (4+x^2)^{-\frac{1}{2}}$$

$$= -x^2(4+x^2)^{-\frac{3}{2}} + (4+x^2)^{-\frac{1}{2}}$$

$$\frac{d^2y}{dx^2} = \frac{-x^2}{(4+x^2)^{\frac{3}{2}}} + \frac{1}{(4+x^2)^{\frac{1}{2}}}$$

Let us now simplify $(4+x^2)\dfrac{d^2y}{dx^2} + x\dfrac{dy}{dx} - y$ as follows:

$$(4+x^2)\left[\frac{-x^2}{(4+x^2)^{\frac{3}{2}}} + \frac{1}{(4+x^2)^{\frac{1}{2}}}\right] + x[1 + x(4+x^2)^{-\frac{1}{2}}] - [x + (4+x^2)^{\frac{1}{2}}]$$

Expanding the brackets gives:

$$-x^2(4+x^2)^{1-\frac{3}{2}} + (4+x^2)^{1-\frac{1}{2}} + x + x^2(4+x^2)^{-\frac{1}{2}} - x - (4+x^2)^{\frac{1}{2}}$$

$$= -x^2(4+x^2)^{-\frac{1}{2}} + (4+x^2)^{\frac{1}{2}} + x + x^2(4+x^2)^{-\frac{1}{2}} - x - (4+x^2)^{\frac{1}{2}}$$

$$= 0 \quad \text{(Note that equal terms with opposite signs cancel out one another to give zero)}$$

Therefore, $(4+x^2)\dfrac{d^2y}{dx^2} + x\dfrac{dy}{dx} - y = 0$ (As proven above)

Exercise 27

1. Find the first, second and third derivatives of $x^4 - 2x^3 - 7x^2 + 1$

2. If $y = \ln 2x^3$, find $\dfrac{d^2y}{dx^2}$.

3. Find $\dfrac{d^2y}{dx^2}$ given that $y = e^{5x^4}$

4. If $y = \sin^2 x^5$, find the second derivative of y.

5. Find $\dfrac{d^3y}{dx^3}$ given that $y = \ln(4x - 10)$

6. Find the second derivative of $y = \dfrac{\sin x}{x}$

7. Find the third derivative of $y = (\ln x)^2$

8. If $y = \cot x$, find $\dfrac{d^3y}{dx^3}$.

9. Given that $y = \sin^2 x + \cos 2x$, find $\dfrac{d^2y}{dx^2} - y$

10. If $y = 2e^{2x} + 5e^{-x}$ evaluate $\dfrac{d^2y}{dx^2} - \dfrac{dy}{dx} - 2y$

11. Given that $y = \cos^2 x + 2\sin x^2$ find $\dfrac{d^2y}{dx^2} - \dfrac{dy}{dx}$

12. Find the second derivative of $y = \dfrac{\tan x}{3x}$

13. If $y = e^{-x}\cos 2x$, find the second derivative of y.

14. If $y = 2\cos 3x + 5\sin 3x$, evaluate $\dfrac{d^2y}{dx^2} + 9y$

15. Find the second derivative of $y = \dfrac{\sin 3x}{3x}$

CHAPTER 28
MISCELLANEOUS PROBLEMS ON DIFFERENTIAL CALCULUS

This chapter covers worked examples on general problems involving combination of topics treated in the various previous chapters. More challenging problems would also be covered here.

Examples

1. If $y = \dfrac{1}{(2x^3 - 5)^4}$ find $\dfrac{dy}{dx}$.

Solution

$$y = \frac{1}{(2x^3 - 5)^4}$$

This can be written as:

$$y = (2x^3 - 5)^{-4}$$

We now use chain rule to differentiate it as follows:

$$\frac{dy}{dx} = -4(2x^3 - 5)^{-4-1} \times \frac{d(2x^3 - 5)}{dx}$$

$$= -4(2x^3 - 5)^{-5} \times 6x^2$$

$$= -24x^2(2x^3 - 5)^{-5}$$

$$\frac{dy}{dx} = \frac{-24x^2}{(2x^3 - 5)^5}$$

2. Given that $f(x) = 5x^4 - 3x^3 - 7x^2 + 9$, find $f'(2)$

Solution

$$f(x) = 5x^4 - 3x^3 - 7x^2 + 9$$

We differentiate $f(x)$ to obtain $f'(x)$ as follows:

$$f'(x) = 20x^3 - 9x^2 - 14x$$

In order to find $f'(2)$, we simply substitute 2 for x in $f'(x)$ as follows:

$$f'(x) = 20x^3 - 9x^2 - 14x$$

$$f'(2) = 20(2^3) - 9(2^2) - 14(2)$$

$$= 20(8) - 9(4) - 28$$

$$= 160 - 36 - 28$$

$$f'(2) = 96$$

3. A function $f(x)$ is given by $f(x) = \dfrac{\sqrt{4 + 3x^2}}{x^3}$. Find:

(a) the derivative of $f(x)$

(b) the gradient of $f(x)$ at the point $(2, \dfrac{1}{2})$

Solution

(a) $f(x) = \dfrac{\sqrt{4 + 3x^2}}{x^3}$

We apply quotient rule to differentiate f(x) as follows:

$$f'(x) = \frac{x^3\left[\dfrac{d(\sqrt{4 + 3x^2})}{dx}\right] - \sqrt{4 + 3x^2}\left[\dfrac{d(x^3)}{dx}\right]}{(x^3)^2}$$

$$= \frac{x^3\left[\dfrac{d(4 + 3x^2)^{\frac{1}{2}}}{dx}\right] - (4 + 3x^2)^{\frac{1}{2}}(3x^2)}{x^6}$$

$$= \frac{x^3\left[\dfrac{1}{2}(4 + 3x^2)^{\frac{1}{2}-1} \times \dfrac{d(3x^2)}{dx}\right] - 3x^2(4 + 3x^2)^{\frac{1}{2}}}{x^6}$$

$$= \frac{x^3\left[\dfrac{1}{2}(4 + 3x^2)^{-\frac{1}{2}} \times 6x\right] - 3x^2(4 + 3x^2)^{\frac{1}{2}}}{x^6}$$

$$= \frac{x^3\left[3x\,(4 + 3x^2)^{-\frac{1}{2}}\right] - 3x^2(4 + 3x^2)^{\frac{1}{2}}}{x^6}$$

$$= \frac{3x^4(4 + 3x^2)^{-\frac{1}{2}} - 3x^2(4 + 3x^2)^{\frac{1}{2}}}{x^6}$$

Let us factorize the expression above by taking $3x^2(4 + 3x^2)^{-\frac{1}{2}}$ as the common factor. This gives:

$$f'(x) = \frac{3x^2(4 + 3x^2)^{-\frac{1}{2}}[x^2 - (4 + 3x^2)]}{x^6}$$

Note that $(4 + 3x^2)^{\frac{1}{2}} \div (4 + 3x^2)^{-\frac{1}{2}} = (4 + 3x^2)^{\frac{1}{2}-(-\frac{1}{2})} = 4 + 3x^2$ as the exponent becomes 1. Recall that exponents are subtracted during division.

$$= \frac{3x^2(4 + 3x^2)^{-\frac{1}{2}}(x^2 - 4 - 3x^2)}{x^6}$$

$$= \frac{3x^2(4 + 3x^2)^{-\frac{1}{2}}(-2x^2 - 4)}{x^6}$$

$$= \frac{-3x^2(4 + 3x^2)^{-\frac{1}{2}}(2x^2 + 4)}{x^6}$$

$$= \frac{-3(4 + 3x^2)^{-\frac{1}{2}}(2x^2 + 4)}{x^4} \qquad \text{(Note that } x^2 \text{ has cancelled out of } x^6\text{)}$$

$$f'(x) = \frac{-3(2x^2 + 4)}{x^4(4 + 3x^2)^{\frac{1}{2}}}$$

(b) The gradient of f(x) at the point $(2, \frac{1}{2})$ is obtained by substituting 2 for x in f'(x). Note that the point $(2, \frac{1}{2})$ is at $x = 2$ and $y = \frac{1}{2}$. We ignore $y = \frac{1}{2}$ since y is not in the expression for f'(x).

Hence, $f'(x) = \dfrac{-3(2x^2 + 4)}{x^4(4 + 3x^2)^{\frac{1}{2}}}$

At $(2, \frac{1}{2})$, $f'(2) = \dfrac{-3(2(2)^2 + 4)}{(2)^4(4 + 3(2)^2)^{\frac{1}{2}}}$

$= \dfrac{-3(8 + 4)}{16(4 + 12)^{\frac{1}{2}}}$

$= \dfrac{-3(12)}{16(16)^{\frac{1}{2}}}$

$= \dfrac{-36}{16 \times 4}$　　(Note that $(16)^{\frac{1}{2}} = \sqrt{16} = 4$)

$f'(2) = \dfrac{-9}{16}$　　(In its lowest term)

Therefore the gradient of f(x) at the point $(2, \frac{1}{2})$ is $\dfrac{-9}{16}$

4. If $x^2y^2 - 3xy + 4xy^3 = 4$, find:

(a) the derivative of the expression

(b) the gradient at (−1, 2)

Solution

(a)　$x^2y^2 - 3xy + 4xy^3 = 4$

The use of implicit differentiation combined with product rule gives us the derivative as follows:

$$x^2\frac{d(y^2)}{dx} + y^2\frac{d(x^2)}{dx} - \left[3x\frac{d(y)}{dx} + y\frac{d(3x)}{dx}\right] + 4x\frac{d(y^3)}{dx} + y^3(4) = \frac{d(4)}{dx}$$

$$x^2\left(2y\frac{dy}{dx}\right) + y^2(2x) - \left[3x\frac{dy}{dx} + y(3)\right] + 4x\left(3y^2\frac{dy}{dx}\right) + 4y^3 = 0$$

$$2x^2y\frac{dy}{dx} + 2xy^2 - 3x\frac{dy}{dx} - 3y + 12xy^2\frac{dy}{dx} + 4y^3 = 0$$

Collect terms in $\frac{dy}{dx}$ on one side of the equation. This gives:

$$2x^2y\frac{dy}{dx} - 3x\frac{dy}{dx} + 12xy^2\frac{dy}{dx} = 3y - 2xy^2 - 4y^3$$

Factorizing the left hand side gives:

$$\frac{dy}{dx}(2x^2y - 3x + 12xy^2) = 3y - 2xy^2 - 4y^3$$

Divide both sides by the terms in the bracket. This gives:

$$\frac{dy}{dx} = \frac{3y - 2xy^2 - 4y^3}{12x^2y - 3x + 12xy^2}$$

(b) At the point (−1, 2) the gradient of the expression is obtained by simply substituting −1 for x and 2 for y in the expression for the derivative. This is done as follows:

$$\frac{dy}{dx} = \frac{3y - 2xy^2 - 4y^3}{12x^2y - 3x + 12xy^2}$$

$$= \frac{3(2) - 2(-1)(2)^2 - 4(2)^3}{12(-1)^2(2) - 3(-1) + 12(-1)(2)^2}$$

$$= \frac{6 + 8 - 32}{24 + 3 - 48}$$

$$= \frac{-18}{-21}$$

$$= \frac{6}{7}$$

5. Find the derivative of $(x + 3y^2)^3 = 7$

Solution

$$(x + 3y^2)^3 = 7$$

We differentiate implicitly and apply chain rule as follows:

$$3(x + 3y^2)^{3-1} \times \frac{d(x+3y^2)}{dx} = 0$$

$$3(x + 3y^2)^2\left(1 + 6y\frac{dy}{dx}\right) = 0$$

Dividing both sides by $3(x + 3y^2)^2$ gives:

$$1 + 6y\frac{dy}{dx} = 0 \qquad \text{(Note that } \frac{0}{3(x + 3y^2)^2} = 0\text{)}$$

$$6y\frac{dy}{dx} = -1$$

$$\frac{dy}{dx} = -\frac{1}{6y}$$

6. If $y = (3x^2 - 2x + 5)(2x - 3)$, find $\frac{dy}{dx}$.

Solution

$$y = (3x^2 - 2x + 5)(2x - 3)$$

We differentiate the expression by applying product rule as follows:

$$\frac{dy}{dx} = (3x^2 - 2x + 5)(2) + (2x - 3)(6x - 2)$$

$$= 6x^2 - 4x + 10 + 12x^2 - 4x - 18x + 6$$

$$\frac{dy}{dx} = 18x^2 - 26x + 16$$

7. Find the derivative of $a^{1 + \tan x}$

Solution

$$y = a^{1 + \tan x}$$

289

Let u = 1 + tan$x$

Hence, y = a^u

$$\frac{du}{dx} = \sec^2 x$$

$$\frac{dy}{du} = a^u \log_e a$$

Hence, $\frac{dy}{dx} = \frac{dy}{du} \times \frac{du}{dx}$

$$= a^u \log_e a \times \sec^2 x$$

$$\frac{dy}{dx} = a^{1+\tan x} \sec^2 x \log_e a$$

8. Find the derivative of $\frac{\log_e x}{1+\cos x}$

Solution

$$y = \frac{\log_e x}{1+\cos x}$$

We apply product rule as follows:

$$\frac{dy}{dx} = \frac{(1+\cos x)\frac{d(\log_e x)}{dx} - \log_e x \frac{d(1+\cos x)}{dx}}{(1+\cos x)^2}$$

$$= \frac{(1+\cos x)\frac{1}{x} - \log_e x(-\sin x)}{(1+\cos x)^2}$$

$$= \frac{\frac{1+\cos x}{x} + \sin x \log_e x}{(1+\cos x)^2}$$

$$= \frac{\frac{1+\cos x + x\sin x\log_e x}{x}}{(1+\cos x)^2}$$

$$\frac{dy}{dx} = \frac{1+\cos x + x\sin x\log_e x}{x(1+\cos x)^2}$$

9. Differentiate with respect to x: ln(cos$x$ + sin$x$)

Solution

$$y = \ln(\cos x + \sin x)$$

$$\frac{dy}{dx} = \frac{\frac{d(\cos x + \sin x)}{dx}}{\cos x + \sin x}$$

$$= \frac{-\sin x + \cos x}{\cos x + \sin x}$$

$$\frac{dy}{dx} = \frac{\cos x - \sin x}{\cos x + \sin x}$$

10. If $y = e^{\sin 2x + \cos x}$ find $\frac{dy}{dx}$.

Solution

$y = e^{\sin 2x + \cos x}$

$$\frac{dy}{dx} = \frac{d(\sin 2x + \cos x)}{dx} \times e^{\sin 2x + \cos x}$$

$\qquad = (2\cos 2x - \sin x)(e^{\sin 2x + \cos x})$

11. Given that $y = e^x - e^{-x}$ show that $\frac{d^3y}{dx^3} + \frac{d^2y}{dx^2} + \frac{dy}{dx} + y = 4e^x$

Solution

$y = e^x - e^{-x}$

$\frac{dy}{dx} = e^x - (-e^{-x})$ (Note that $\frac{d(e^{-x})}{dx} = \frac{d(-x)}{dx} \times e^{-x} = -1 \times e^{-x} = -e^{-x}$)

$\qquad = e^x + e^{-x}$

$\frac{d^2y}{dx^2} = e^x - e^{-x}$

$\frac{d^3y}{dx^3} = e^x + e^{-x}$

Let us now substitute corresponding term into $\frac{d^3y}{dx^3} + \frac{d^2y}{dx^2} + \frac{dy}{dx} + y$ as follows:

$(e^x + e^{-x}) + (e^x - e^{-x}) + (e^x + e^{-x}) + (e^x - e^{-x})$

Collecting like terms together gives:

$e^x + e^x + e^x + e^x + e^{-x} - e^{-x} + e^{-x} - e^{-x}$

$= 4e^x$ (Note that e^{-x} cancels out each other)

Hence, $\frac{d^3y}{dx^3} + \frac{d^2y}{dx^2} + \frac{dy}{dx} + y = 4e^x$ (As proven above)

12. Differentiate with respect to x: $\ln\left(\frac{1 - 3x^2}{1 + 3x^2}\right)^{\frac{1}{2}}$

Solution

$y = \ln\left(\frac{1 - 3x^2}{1 + 3x^2}\right)^{\frac{1}{2}}$

$\qquad = \log_e\left(\frac{1 - 3x^2}{1 + 3x^2}\right)^{\frac{1}{2}}$ (Note that "ln" is $\log_e$)

$\qquad = \log_e\left[\frac{(1 - 3x^2)^{\frac{1}{2}}}{(1 + 3x^2)^{\frac{1}{2}}}\right]$

$$= \log_e(1 - 3x^2)^{\frac{1}{2}} - \log_e(1 + 3x^2)^{\frac{1}{2}}$$

$$y = \frac{1}{2} \log_e(1 - 3x^2) - \frac{1}{2} \log_e(1 + 3x^2)$$

$$\frac{dy}{dx} = \frac{1}{2} \frac{\frac{d(1-3x^2)}{dx}}{1-3x^2} - \frac{1}{2} \frac{\frac{d(1+3x^2)}{dx}}{1+3x^2}$$

$$= \frac{1}{2} \frac{-6x}{1-3x^2} - \frac{1}{2} \frac{6x}{1+3x^2}$$

$$= \frac{-3x}{1-3x^2} - \frac{3x}{1+3x^2} \qquad \text{(Note that } \frac{1}{2} \text{ reduces } 6x \text{ to } 3x)$$

$$= \frac{-3x(1+3x^2) - 3x(1-3x^2)}{(1-3x^2)(1+3x^2)}$$

$$= \frac{-3x - 9x^3 - 3x + 9x^3)}{(1-3x^2)(1+3x^2)}$$

$$= \frac{-6x}{1 + 3x^2 - 3x^2 - 9x^4}$$

$$\frac{dy}{dx} = \frac{-6x}{1 - 9x^4}$$

13. If $y = \frac{x}{\sqrt{9 - x^2}}$ show that: $(9 - x^2)\frac{d^2y}{dx^2} = 3x\frac{dy}{dx}$

Solution

$$y = \frac{x}{\sqrt{9 - x^2}}$$

Applying quotient rule gives $\frac{dy}{dx}$ as follows:

$$\frac{dy}{dx} = \frac{\sqrt{9 - x^2}(1) - x\frac{d(\sqrt{9 - x^2})}{dx}}{(\sqrt{9 - x^2})^2}$$

$$= \frac{(9 - x^2)^{\frac{1}{2}}(1) - x\frac{d[(9 - x^2)^{\frac{1}{2}}]}{dx}}{[(9 - x^2)^{\frac{1}{2}}]^2}$$

$$= \frac{(9 - x^2)^{\frac{1}{2}} - x\left[\frac{1}{2}(9 - x^2)^{\frac{1}{2} - 1}\right] \times (-2x)}{9 - x^2} \qquad \text{(Note that } -2x \text{ is from the derivative of } -x^2)$$

$$= \frac{(9 - x^2)^{\frac{1}{2}} - x\left[-x(9 - x^2)^{-\frac{1}{2}}\right]}{9 - x^2}$$

$$= \frac{(9 - x^2)^{\frac{1}{2}} + x^2(9 - x^2)^{-\frac{1}{2}}}{9 - x^2}$$

$$= \frac{(9 - x^2)^{\frac{1}{2}} + \frac{x^2}{(9 - x^2)^{\frac{1}{2}}}}{9 - x^2}$$

$$= \frac{\frac{9 - x^2 + x^2}{(9 - x^2)^{\frac{1}{2}}}}{9 - x^2}$$

$$= \frac{9}{(9 - x^2)^{\frac{1}{2}}(9 - x^2)}$$

$$= \frac{9}{(9 - x^2)^{\frac{3}{2}}}$$

$$\frac{dy}{dx} = 9(9 - x^2)^{-\frac{3}{2}}$$

Applying chain rule gives $\frac{d^2y}{dx^2}$ as follows:

$$\frac{d^2y}{dx^2} = \frac{-3}{2} \times 9(9 - x^2)^{\frac{3}{2} - 1} \times -2x \qquad \text{(Note that } -2x \text{ is from the derivative of } -x^2)$$

$$= \frac{-27}{2} \times -2x(9 - x^2)^{-\frac{5}{2}}$$

$$= 27x(9 - x^2)^{-\frac{5}{2}}$$

$$\frac{d^2y}{dx^2} = \frac{27x}{(9 - x^2)^{\frac{5}{2}}}$$

Let us now show that $(9 - x^2)\frac{d^2y}{dx^2} = 3x\frac{dy}{dx}$

We simplify $(9 - x^2)\frac{d^2y}{dx^2}$ as follows:

$$(9 - x^2)\frac{27x}{(9 - x^2)^{\frac{5}{2}}}$$

$$= \frac{27x(9 - x^2)}{(9 - x^2)^{\frac{5}{2}}}$$

$$= 27x(9 - x^2)^{1 - \frac{5}{2}}$$

$$= 27x(9 - x^2)^{-\frac{3}{2}}$$

$$= \frac{27x}{(9 - x^2)^{\frac{3}{2}}}$$

Hence $(9 - x^2)\frac{d^2y}{dx^2}$ gives us $\dfrac{27x}{(9 - x^2)^{\frac{3}{2}}}$

Let us now simplify $3x\frac{dy}{dx}$ as follows:

$$3x[9(9 - x^2)^{-\frac{3}{2}}] \qquad \text{(Note that } \frac{dy}{dx} = 9(9 - x^2)^{-\frac{3}{2}} \text{ as obtained above)}$$

$$= 27x(9 - x^2)^{-\frac{3}{2}}$$

$$= \frac{27x}{(9 - x^2)^{\frac{3}{2}}}$$

Hence $3x\dfrac{dy}{dx}$ also gives us $\dfrac{27x}{(9-x^2)^{\frac{3}{2}}}$

Therefore, $(9-x^2)\dfrac{d^2y}{dx^2} = 3x\dfrac{dy}{dx}$ as both sides give $\dfrac{27x}{(9-x^2)^{\frac{3}{2}}}$

14. Given that $y = \dfrac{x}{x-1}$, show that: $(x-1)\dfrac{d^2y}{dx^2} + 2\dfrac{dy}{dx} = 0$

Solution

$$y = \dfrac{x}{x-1}$$

Using quotient rule gives $\dfrac{dy}{dx}$ as follows:

$$\dfrac{dy}{dx} = \dfrac{(x-1)(1) - x(1)}{(x-1)^2}$$

$$= \dfrac{x-1-x}{(x-1)^2}$$

$$\dfrac{dy}{dx} = \dfrac{-1}{(x-1)^2}$$

Using chain rule gives $\dfrac{d^2y}{dx^2}$ as follows:

$$\dfrac{dy}{dx} = \dfrac{-1}{(x-1)^2}$$

$$= -1(x-1)^{-2}$$

$$\dfrac{d^2y}{dx^2} = -2 \times -1(x-1)^{-2-1} \times \dfrac{d(x)}{dx}$$

$$= 2(x-1)^{-3} \times 1$$

$$\dfrac{d^2y}{dx^2} = \dfrac{2}{(x-1)^3}$$

From the question, let us now simplify $(x-1)\dfrac{d^2y}{dx^2} + 2\dfrac{dy}{dx}$ as follows:

$$(x-1)\dfrac{d^2y}{dx^2} + 2\dfrac{dy}{dx}$$

$$= (x-1)\dfrac{2}{(x-1)^3} + (2)\dfrac{-1}{(x-1)^2}$$

$$= \dfrac{2}{(x-1)^2} - \dfrac{2}{(x-1)^2}$$

$$= 0$$

This shows that $(x-1)\dfrac{d^2y}{dx^2} + 2\dfrac{dy}{dx}$ is equal to zero.

15. Find, with respect to x, the derivative of $\left(x - \dfrac{5}{x}\right)^3$

<u>Solution</u>

$$y = \left(x - \frac{5}{x}\right)^3$$

This can also be written as:

$$y = (x - 5x^{-1})^3$$

$$\frac{dy}{dx} = 3(x - 5x^{-1})^{3-1} \times \frac{d(x-5x^{-1})}{dx} \qquad \text{(By use of chain rule)}$$

$$= 3(x - 5x^{-1})^2 \times 1 - (-1 \times 5x^{-1-1})$$

$$= 3(x - 5x^{-1})^2 \times 1 - (-5x^{-2})$$

$$= 3(x - 5x^{-1})^2 \times (1 + 5x^{-2})$$

$$= 3(x - 5x^{-1})^2(1 + 5x^{-2})$$

$$\frac{dy}{dx} = 3\left(x - \frac{5}{x}\right)^2 \left(1 + \frac{5}{x^2}\right)$$

16. Given that $\exp(2x^2 + 2y^2 - 16) = x + y$, find:

(a) $\dfrac{dy}{dx}$

(b) $\dfrac{dy}{dx}$ at $\left(\dfrac{1}{2}, \dfrac{1}{2}\right)$

<u>Solution</u>

$$\exp(2x^2 + 2y^2 - 16) = x + y$$

This can also be written as

$$e^{2x^2 + 2y^2 - 16} = x + y \qquad \text{(Note that } \exp x = e^x\text{)}$$

We now differentiate implicitly as follows:

$$\frac{d(2x^2 + 2y^2 - 16)}{dx} \times (e^{2x^2 + 2y^2 - 16}) = \frac{d(x)}{dx} + \frac{d(y)}{dx}$$

$$\left(4x + 4y\frac{dy}{dx}\right)(e^{2x^2 + 2y^2 - 16}) = 1 + \frac{dy}{dx}$$

Expanding the bracket gives:

$$4x(e^{2x^2 + 2y^2 - 16}) + 4y\frac{dy}{dx}(e^{2x^2 + 2y^2 - 16}) = 1 + \frac{dy}{dx}$$

Collecting terms in $\dfrac{dy}{dx}$ on the left hand side gives:

$$4y\frac{dy}{dx}(e^{2x^2 + 2y^2 - 16}) - \frac{dy}{dx} = 1 - 4x(e^{2x^2 + 2y^2 - 16})$$

Factorizing the left hand side gives:

$$\frac{dy}{dx}[4y(e^{2x^2 + 2y^2 - 16}) - 1] = 1 - 4x(e^{2x^2 + 2y^2 - 16})$$

Hence, $\dfrac{dy}{dx} = \dfrac{1 - 4x(e^{2x^2 + 2y^2 - 16})}{4y(e^{2x^2 + 2y^2 - 16}) - 1}$

Or, $\dfrac{dy}{dx} = \dfrac{-[4x(e^{2x^2 + 2y^2 - 16}) - 1]}{4y(e^{2x^2 + 2y^2 - 16}) - 1}$

(b) In order to find $\dfrac{dy}{dx}$ at $\left(\dfrac{1}{2}\ \dfrac{1}{2}\right)$, we simply substitute $x = \dfrac{1}{2}$ and $y = \dfrac{1}{2}$ into the expression for $\dfrac{dy}{dx}$ as follows:

$$\frac{dy}{dx} = \frac{-[4x(e^{2x^2+2y^2-16}) - 1]}{4y(e^{2x^2+2y^2-16}) - 1}$$

Since $x = \dfrac{1}{2}$ and $y = \dfrac{1}{2}$, this simplifies to give:

$$\frac{dy}{dx} \text{ at } \left(\frac{1}{2}\ \frac{1}{2}\right) = \frac{-[4\left(\frac{1}{2}\right)(e^{2x^2+2y^2-16}) - 1]}{4\left(\frac{1}{2}\right)(e^{2x^2+2y^2-16}) - 1}$$

$$= \frac{-[2(e^{2x^2+2y^2-16}) - 1]}{2(e^{2x^2+2y^2-16}) - 1}$$

The numerator cancels out the denominator to give:

$$\frac{dy}{dx} = \frac{-1}{1}$$

$$\frac{dy}{dx} = -1$$

17. If $y = x^3 - 2x^2 + 5$, show that: $x\dfrac{dy}{dx} - 3y - 2x^2 + 15 = 0$

Solution

$$y = x^3 - 2x^2 + 5$$

$$\frac{dy}{dx} = 3x^2 - 4x$$

Let us now simplify $x\dfrac{dy}{dx} - 3y - 2x^2 + 15$ as follows:

$$x\frac{dy}{dx} - 3y - 2x^2 + 15$$

$$= x(3x^2 - 4x) - 3y - 2x^2 + 15$$

$$= x(3x^2 - 4x) - 3(x^3 - 2x^2 + 5) - 2x^2 + 15 \quad \text{(Note that } x^3 - 2x^2 + 5 \text{ has been substituted for y)}$$

$$= 3x^3 - 4x^2 - 3x^3 + 6x^2 - 15 - 2x^2 + 15$$

$$= 0$$

Therefore, $x\dfrac{dy}{dx} - 3y - 2x^2 + 15 = 0$ as proven above.

18. Determine $\dfrac{d^2}{dx^2}\left(x\sin\dfrac{1}{x}\right)$

Solution

This means the second derivative of $x\sin\dfrac{1}{x}$

Let $y = x\sin\dfrac{1}{x}$

Or, $y = x\sin x^{-1}$

We now apply product rule to obtain $\dfrac{dy}{dx}$ as follows:

$$\frac{dy}{dx} = x\left[\frac{d(x^{-1})}{dx}\cos x^{-1}\right] + \sin x^{-1}\left[\frac{d(x)}{dx}\right]$$

$$= x(-x^{-2}\cos x^{-1}) + \sin x^{-1}(1)$$

$$= x(\frac{-1}{x^2}\cos x^{-1}) + \sin x^{-1}$$

$$= \frac{-x}{x^2}\cos x^{-1} + \sin x^{-1}$$

$$= \frac{-\cos x^{-1}}{x} + \sin x^{-1}$$

$$\frac{dy}{dx} = -\frac{1}{x}\cos\frac{1}{x} + \sin\frac{1}{x}$$

Or, $\dfrac{dy}{dx} = -x^{-1}\cos x^{-1} + \sin x^{-1}$

We now obtain $\dfrac{d^2y}{dx^2}$ as follows:

$$\frac{d^2y}{dx^2} = -x^{-1}\left[\frac{d(\cos x^{-1})}{dx}\right] + \cos x^{-1}\left[\frac{d(-x^{-1})}{dx}\right] + \frac{d(\sin x^{-1})}{dx}$$

$$= -x^{-1}\left[\frac{d(x^{-1})}{dx}(-\sin x^{-1})\right] + \cos x^{-1}(x^{-2}) + \left[\frac{d(x^{-1})}{dx}\cos x^{-1}\right]$$

$$= -x^{-1}[-x^{-2}(-\sin x^{-1})] + \cos x^{-1}(x^{-2}) - x^{-2}(\cos x^{-1})$$

$$= -x^{-1}(x^{-2}\sin x^{-1}) + x^{-2}\cos x^{-1} - x^{-2}\cos x^{-1}$$

$$= -x^{-1}(x^{-2}\sin x^{-1}) \quad \text{(Note that } x^{-2}\cos x^{-1} \text{ cancels out)}$$

$$= -x^{-3}\sin x^{-1}$$

$$\frac{d^2y}{dx^2} = -\frac{1}{x^3}\sin\frac{1}{x}$$

19. Given that $y = \dfrac{e^x + e^{-x}}{e^x - e^{-x}}$

(a) find $\dfrac{d^2y}{dx^2}$

(b) show that $\dfrac{d^2y}{dx^2} + 2y\dfrac{dy}{dx} = 0$

Solution

(a) $y = \dfrac{e^x + e^{-x}}{e^x - e^{-x}}$

By using quotient rule we obtain $\dfrac{dy}{dx}$ as follows:

$$\frac{dy}{dx} = \frac{e^x - e^{-x}\left[\frac{d(e^x + e^{-x})}{dx}\right] - \left(e^x + e^{-x}\left[\frac{d(e^x - e^{-x})}{dx}\right]\right)}{(e^x - e^{-x})^2}$$

$$= \frac{(e^x - e^{-x})(e^x - e^{-x}) - (e^x + e^{-x})(e^x + e^{-x})}{(e^x - e^{-x})^2}$$

$$= \frac{(e^x - e^{-x})^2 - (e^x + e^{-x})^2}{(e^x - e^{-x})^2}$$

Separating into fractions gives:

$$= \frac{(e^x - e^{-x})^2}{(e^x - e^{-x})^2} - \frac{(e^x + e^{-x})^2}{(e^x - e^{-x})^2}$$

$$\frac{dy}{dx} = 1 - \frac{(e^x + e^{-x})^2}{(e^x - e^{-x})^2}$$

We now apply quotient and chain rules to obtain $\dfrac{d^2y}{dx^2}$ as follows:

$$\frac{d^2y}{dx^2} = 0 - \frac{(e^x - e^{-x})^2[2(e^x + e^{-x})(e^x - e^{-x})] - (e^x + e^{-x})^2[2(e^x - e^{-x})(e^x + e^{-x})]}{[(e^x - e^{-x})^2]^2}$$

$$= -\left[\frac{2(e^x - e^{-x})^3(e^x + e^{-x}) - 2(e^x + e^{-x})^3(e^x - e^{-x})}{(e^x - e^{-x})^4}\right]$$

$$= \frac{-2(e^x - e^{-x})^3(e^x + e^{-x}) + 2(e^x + e^{-x})^3(e^x - e^{-x})}{(e^x - e^{-x})^4}$$

$$= \frac{2(e^x + e^{-x})^3(e^x - e^{-x}) - 2(e^x - e^{-x})^3(e^x + e^{-x})}{(e^x - e^{-x})^4} \qquad \text{(After rearranging the numerator)}$$

Separating into fractions gives:

$$= \frac{2(e^x + e^{-x})^3(e^x - e^{-x})}{(e^x - e^{-x})^4} - \frac{2(e^x - e^{-x})^3(e^x + e^{-x})}{(e^x - e^{-x})^4}$$

$$\frac{d^2y}{dx^2} = \frac{2(e^x + e^{-x})^3}{(e^x - e^{-x})^3} - \frac{2(e^x + e^{-x})}{e^x - e^{-x}}$$

(b) Let us simplify $\dfrac{d^2y}{dx^2} + 2y\dfrac{dy}{dx}$ as follows:

$$\frac{d^2y}{dx^2} + 2y\frac{dy}{dx}$$

$$= \frac{2(e^x + e^{-x})^3}{(e^x - e^{-x})^3} - \frac{2(e^x + e^{-x})}{e^x - e^{-x}} + 2\left(\frac{e^x + e^{-x}}{e^x - e^{-x}}\right)\left[1 - \frac{(e^x + e^{-x})^2}{(e^x - e^{-x})^2}\right]$$

Expanding bracket gives:

$$= \frac{2(e^x + e^{-x})^3}{(e^x - e^{-x})^3} - \frac{2(e^x + e^{-x})}{e^x - e^{-x}} + 2\left(\frac{e^x + e^{-x}}{e^x - e^{-x}}\right) - \frac{2(e^x + e^{-x})^3}{(e^x - e^{-x})^3}$$

$$= 0 \qquad \text{(Since equal terms with opposite signs cancel out each other)}$$

Therefore, $\dfrac{d^2y}{dx^2} + 2y\dfrac{dy}{dx}$ gives zero as proven above.

Or, $\dfrac{d^2y}{dx^2} + 2y\dfrac{dy}{dx} = 0$

20. Find the derivative of In(tan2x)

Solution

$y = \text{In}(\tan 2x)$

$$\frac{dy}{dx} = \frac{\frac{d(\tan 2x)}{dx}}{\tan 2x}$$

$$= \frac{2\sec^2 2x}{\tan 2x}$$

Or, $\dfrac{dy}{dx} = \dfrac{2\left(\frac{1}{\cos 2x}\right)\left(\frac{1}{\cos 2x}\right)}{\left(\frac{\sin 2x}{\cos 2x}\right)}$ (Note that $\sec 2x = \dfrac{1}{\cos 2x}$ and $\tan 2x = \dfrac{\sin 2x}{\cos 2x}$)

$$= \left(\frac{2}{\cos 2x}\right)\left(\frac{1}{\cos 2x}\right) \times \frac{\cos 2x}{\sin 2x}$$

$$\frac{dy}{dx} = \frac{2}{\sin 2x \cos 2x} \quad \text{(Since } \cos 2x \text{ cancels } \cos 2x\text{)}$$

21. Given that $y = x^3 + 3x^2$, determine $2\dfrac{dy}{dx} - x\dfrac{d^2y}{dx^2}$

Solution

$$y = x^3 + 3x^2$$

$$\frac{dy}{dx} = 3x^2 + 6x$$

$$\frac{d^2y}{dx^2} = 6x + 6$$

Hence, $2\dfrac{dy}{dx} - x\dfrac{d^2y}{dx^2}$ is simplified as follows:

$$2(3x^2 + 6x) - x(6x + 6)$$

$$= 6x^2 + 12x - 6x^2 - 6x$$

$$= 6x$$

Therefore, $2\dfrac{dy}{dx} - x\dfrac{d^2y}{dx^2} = 6x$

22. Find $\dfrac{dy}{dx}$ if $y = 3(3x + \sqrt{x})^2$

Solution

$$y = 3(3x + \sqrt{x})^2$$

Applying chain rule gives:

$$\frac{dy}{dx} = 3 \times 2\,(3x + x^{\frac{1}{2}})^{2-1} \times \frac{d(3x + x^{\frac{1}{2}})}{dx} \quad \text{(Note that } \sqrt{x} = x^{\frac{1}{2}}\text{)}$$

$$= 6(3x + x^{\frac{1}{2}}) \times (3 + \frac{1}{2}x^{-\frac{1}{2}})$$

$$= 6(3x + x^{\frac{1}{2}})\left(3 + \frac{1}{2x^{\frac{1}{2}}}\right)$$

$$= 6(3x + \sqrt{x})\left(3 + \frac{1}{2\sqrt{x}}\right)$$

Expanding the bracket gives:

$$= 6\left(9x + \frac{3x}{2\sqrt{x}} + 3\sqrt{x} + \frac{1}{2}\right)$$

$$= 6\left(9x + \frac{3}{2}\sqrt{x} + 3\sqrt{x} + \frac{1}{2}\right) \qquad \left(\text{Note that } \frac{3x}{2\sqrt{x}} = \frac{3}{2}x^{1-\frac{1}{2}} = \frac{3}{2}x^{\frac{1}{2}} = 3\sqrt{x}\right)$$

$$= 6\left(9x + \frac{9}{2}\sqrt{x} + \frac{1}{2}\right)$$

$$\frac{dy}{dx} = 54x + 27\sqrt{x} + 3$$

23. If $y = \dfrac{5x^2 + 7}{x^4}$,

(a) find $\dfrac{d^2y}{dx^2}$

(b) show that $x^2\dfrac{d^2y}{dx^2} + 7x\dfrac{dy}{dx} + 8y = 0$

Solution

(a) $y = \dfrac{5x^2 + 7}{x^4}$

$$= \frac{5x^2}{x^4} + \frac{7}{x^4}$$

$$= \frac{5}{x^2} + \frac{7}{x^4}$$

$$y = 5x^{-2} + 7x^{-4}$$

$$\frac{dy}{dx} = -10x^{-3} - 28x^{-5}$$

Similarly,

$$\frac{d^2y}{dx^2} = 30x^{-4} + 140x^{-6}$$

(b) Let us now simplify $x^2\dfrac{d^2y}{dx^2} + 7x\dfrac{dy}{dx} + 8y$ by substituting appropriately as follows:

$$x^2(30x^{-4} + 140x^{-6}) + 7x(-10x^{-3} - 28x^{-5}) + 8(5x^{-2} + 7x^{-4})$$

$$= 30x^{-2} + 140x^{-4} - 70x^{-2} - 196x^{-4} + 40x^{-2} + 56x^{-4}$$

$$= 30x^{-2} + 40x^{-2} - 70x^{-2} + 140x^{-4} + 56x^{-4} - 196x^{-4}$$

$$= 0$$

Therefore $x^2\dfrac{d^2y}{dx^2} + 7x\dfrac{dy}{dx} + 8y$ gives zero as obtained above.

Or, $x^2\dfrac{d^2y}{dx^2} + 7x\dfrac{dy}{dx} + 8y = 0$

24. If $y = (2x + 5)^4 + \dfrac{x - 1}{2x - 1}$ find $\dfrac{dy}{dx}$.

Solution

$$(2x + 5)^4 + \frac{x - 1}{2x - 1}$$

We now use chain rule and quotient rule as follows:

$$\frac{dy}{dx} = 4(2x+5)^{4-1} \times \frac{d(2x+5)}{dx} + \frac{(2x-1)\frac{d(x-1)}{dx} - \left[(x-1)\frac{d(2x-1)}{dx}\right]}{(2x-1)^2}$$

$$= 4(2x+5)^3(2) + \frac{(2x-1)(1) - [(x-1)(2)]}{(2x-1)^2}$$

$$= 8(2x+5)^3 + \frac{(2x-1) - 2(x-1)}{(2x-1)^2}$$

$$= 8(2x+5)^3 + \frac{2x-1-2x+2}{(2x-1)^2}$$

$$\frac{dy}{dx} = 8(2x+5)^3 + \frac{1}{(2x-1)^2}$$

Exercise 28

1. If $y = x^2(3x^4 - 5)^3$ find $\frac{dy}{dx}$.

2. Given that $f(x) = 2x^5 - x^4 + 2x^3 + x^2 - 3x + 4$ find $f'(-2)$

3. A function $f(x)$ is given by $f(x) = \frac{\sqrt{(3x^2 - 1)^3}}{x^2}$. Find:

(a) the derivative of $f(x)$
(b) the gradient of $f(x)$ at the point $(1, -2)$

4. If $3xy^3 - y^2 - 4x^3 = 5y$, find:

(a) the derivative of the expression
(b) the gradient at $(-1, 1)$

5. Find the derivative of $(2x^2 + y^2)^2 = 0$

6. If $y = (x^2 + 5)^3(x^3 - 1)$, find $\frac{dy}{dx}$.

7. Find the derivative of $e^{\sin x + \tan x}$

8. Find the derivative of $\frac{x^2}{\sin^2 x}$

9. Differentiate with respect to x: $\ln(\sin^2 x + \cos 3x)$

10. If $y = a^{\cos 5x}$ find $\frac{dy}{dx}$.

11. Given that $y = x^2 e^{-3x}$ evaluate $\frac{d^2y}{dx^2} + \frac{dy}{dx} - y$

12. Differentiate with respect to x: $\sin\left(\frac{1 - 2x^3}{x^2}\right)$

13. If $y = \frac{2x - 1}{x^2}$, find $x^4\frac{d^2y}{dx^2} - 3(2x - 1)\frac{dy}{dx}$

301

14. Given that $y = 5e^{-2x} + 3e^x$, evaluate $\dfrac{d^2y}{dx^2} + \dfrac{dy}{dx} - 2y$

15. Find the derivative of $\left(\dfrac{1}{x^2} - \dfrac{2}{x}\right)^5$

16. Given that $e^{x^3 - y^3} = xy$, find:

(a) $\dfrac{dy}{dx}$

(b) $\dfrac{dy}{dx}$ at $(1, -1)$

17. If $y = \sin(\sin x)$, evaluate $\dfrac{d^2y}{dx^2} + \tan x \dfrac{dy}{dx} + y\cos^2 x$

18. Determine $\dfrac{d^2}{dx^2}\left(x^2\cos\dfrac{1}{x^2}\right)$

19. Given that $y = \dfrac{1 + e^{-x}}{1 - e^{-x}}$

(a) find $\dfrac{d^2y}{dx^2}$

(b) evaluate $\dfrac{d^2y}{dx^2} - \dfrac{dy}{dx} + y$

20. Find the derivative of $\ln(\sec^2 x)$

21. Given that $y = x + \tan x$, determine $\cos^2 x \dfrac{d^2y}{dx^2} - 2y + 2x$

22. Find $\dfrac{dy}{dx}$ if $y = \left(3x + \dfrac{\sqrt{5x}}{2}\right)^3$

23. If $y = \dfrac{x^3 + 2x^2 - 5x - 3}{x^2}$, find $\dfrac{d^2y}{dx^2}$

24. If $y = (x^3 + 1)^3 + \dfrac{5x - 2}{x^2 + 1}$ find $\dfrac{dy}{dx}$.

25. Given that $y = 2\cos 5x + 7\sin 5x$, determine $\dfrac{d^2y}{dx^2} + 25y$

ANSWERS TO EXERCISES

Exercise 1

1. $\dfrac{m+d}{m+f}$
2. $\dfrac{x-7}{x-4}$
3. $\dfrac{2-x}{3x-5}$
4. $-\dfrac{(4m+n)}{(n+m)}$
5. $\dfrac{p-r}{p+r}$
6. $\dfrac{6a+7}{3b}$
7. $\dfrac{13}{5(2x-y)}$
8. $\dfrac{1+9m}{6mn}$

9. $\dfrac{5a^2+13a+4}{(a-2)(a+3)}$
10. $\dfrac{3ab^2+4b^2-3a^3}{12a^2b^3}$
11. $\dfrac{4+5n}{(3m-2n)(3m+2n)}$

Exercise 2

1. $m=-4$
2. $b=-\dfrac{3}{2}$ or 6
3. $x=7$ or 4.875
4. $x=-8$
5. $x=\dfrac{7}{3}$
6 (a) $a=-\dfrac{12}{5}$

(b) $a=6$ or -5
7. (a) $m=-2$
(b) $m=-\dfrac{9}{5}$
8 (a) $x=\dfrac{2}{5}$ or -3
(b) $x=-\dfrac{5}{2}$ or 3
9. $m=\dfrac{22}{5}$

10. $x=-\dfrac{41}{113}$
11. 4
12. $\dfrac{33}{26}$

Exercise 3

1. $x=1, y=2$
2. $a=\dfrac{1}{4}$, $b=-1$
3. $c=2, d=3$
4. $a=\dfrac{4}{3}$, $b=\dfrac{2}{3}$
5. $p=-\dfrac{1}{2}$, $q=\dfrac{1}{2}$

6. $c=2, d=-5$
7. $x=\dfrac{3}{10}$, $y=\dfrac{2}{5}$
8. $m=\dfrac{9}{17}$, $n=9$

Exercise 4

1. $x=9$ or -9
2. (a) $x=2$ or $-\dfrac{24}{5}$
(b) $x=\dfrac{7}{3}$ or -3
(c) $x=-\dfrac{3}{4}$ or $\dfrac{33}{4}$
3 (a) $x=\dfrac{5}{4}$ or -3

(b) No solution
4 (a) $x=\dfrac{7}{2}$ or -2
(b) $x=-5$ or -13
5 (a) $x=2$
(b) $x=35$ or -3

(c) $x=-\dfrac{1}{2}$, or $\dfrac{1}{2}$
6 (a) $x=2$ or -9
(b) $x=0,\dfrac{9}{2}$, $x=\dfrac{1}{2}$ or 4
7 (a) $x=12$
(b) $x=1$

8 (a) No solution
(b) $x=5$ or -2
9 (a) $x=-\dfrac{2}{3}$ or 3
(b) $x=3.63, -17.63, 4$ or -18

10 (a) $x=\dfrac{3}{5}$ or 2
(b) $x=1.30, -5.80, 0.84$ or -5.34

Exercise 5

1. $x\leq-2$ or $x\geq\dfrac{4}{5}$
2. $-5<x<\dfrac{19}{3}$
3. $-3\leq x\leq19$
4. $m<-1$ or $m>-\dfrac{3}{7}$
5. $x<-3$ or $x>$ 5

6. $-3<m<2$
7. $2<m\leq5$
8. $-2<x\leq-1$
9. $x<-3$ or $x>3$
10. $y<6$ or $y>-6$

11. $a\leq-2.1$ or $a\geq5.4$
12. $y\leq-3$ or $y\geq3$
13. $m\leq-1$ or $m\geq1$
14. $x>-\dfrac{1}{2}$ or $x\leq2$

15. $-4\leq y\leq4$

Exercise 6

1. $x = \dfrac{16}{5}$ 2. $x = \dfrac{13}{11}$ 3. $x = \dfrac{14}{13}$ 4. $x = 0.42$ 5. $x = -41.39$ 6. $x = 0.474$ 7. $x = 2.50$ or -0.34

8. $x = 4.14$ or 0.36 9. $x = 3$ 10. $x = 2$ 11. $x = 0$ 12. $x = 1, y = -1$ 13. $x = 5, y = \dfrac{20}{3}$

14. $y = -\dfrac{4}{9}, x = \dfrac{19}{9}$ 15. $x = -\dfrac{4}{3}$ or 2

Execise 7

1 (a) $\dfrac{8}{3}$ (b) $\dfrac{5}{3}$ (c) $\dfrac{34}{9}$ (d) $\dfrac{34}{15}$ (e) $\dfrac{8}{15}$ (f) $\dfrac{34}{25}$ (g) $\dfrac{152}{27}$ 2. $4x^2 - 188x + 121 = 0$

3. $40x^2 + 89x + 40 = 0$ 4. $2x^2 + 15x - 1 = 0$ 5. $49x^2 - 189x + 9 = 0$ 6. $8x^2 + 305x - 216 = 0$

7 (a) $\dfrac{\sqrt{24}}{5}$ (b) $\dfrac{8\sqrt{24}}{25}$ 8. $5\dfrac{9}{20}$ 9. $\dfrac{13\sqrt{13}}{8}$ 10 (a) $3x^2 - 28x + 32 = 0$ (b) $3x^2 - 8x - 16 = 0$

11. $m = -9$ and $n = \dfrac{2}{3}$ or $m = -2$ and $n = 3$ 12. $P = 1$ or -23 13. $k = -\dfrac{28}{3}$ 14. $\dfrac{84}{5}$ 15 (a) $y = -\dfrac{217}{8}$

(b) $x = -\dfrac{11}{4}$ 16 (a) $y = \dfrac{513}{16}$ (b) $x = -\dfrac{15}{8}$ 17 (a) 1 and 3 or -1 and -3 (b) $k = -6$ or 10

18. $3\sqrt{3}$ 19. $K = -8$ 20. $\dfrac{\sqrt{5}}{2}$ and $\sqrt{5}$ or $-\dfrac{\sqrt{5}}{2}$ and $-\sqrt{5}$

Exercise 8

1 (a) It is not a function   (b) It is not a function   (c) It is not a function   (d) It is a function

2 (a) -5 (b) 22   (c) -1.625 (d) $81x^3 - 162x^2 + 108x - 26$ (e) $3x^3 + 27x^2 + 81x - 79$

3 (a) $-\dfrac{13}{5}$ (b) $p = \dfrac{7}{4}$ (c) $-\dfrac{153}{5}$ 4 (a) $3x - 2$ (b) $16 - 3x$ (c) $2x^2 - 5x - 3$ (d) $-\dfrac{4}{5}$

(e) $h(x) = x - 16$ (f) -14 5 (a) $2x - 4$ (b) $2x - 2$ 6 (a) $18(4x^2 + 4x + 1)$ (b) $18x^2$

(c) $36x^2 + 1$ 7 (a) Even   (b) Odd   8 (a) Even   (b) Even   (c) Neither   (d) Even

9 (a) $\dfrac{x + 1}{5}$ (b) $\sqrt[3]{\dfrac{x - 2}{9}}$ (c) $\sqrt[5]{\dfrac{x}{2}}$ 10 (a) $2\sqrt[3]{3} + 1$ (b) $\dfrac{1 + 2x^2}{5x^2 - 2}$ 11 (a) $\dfrac{x + 1}{5 - 2x}$ (b) 0

12 (a) $2x^2 + 3x - 9$ (b) -7 13. $\dfrac{2x - 2}{5} + 7$ 14. 47   15 (a) $2x - 5$ (b) $\dfrac{x + 5}{2}$ (c) -9

(d) 2   (e) $4x - 7$ 16 (a) $7x - 12$ (b) $14x^2 + 23$

Exercise 9

1 (a) $5x^2 + 5x + 6$ (b) $-2x^2 + 4x - 22$ (c) $3x^3 - 8x^2 + 6x - 11$ (d) $3x^3 + 7x^2 - 4x + 21$

(e) $9x^3 + 4x^2 + 2x + 9$ 2 (a) -82 (b) -10 3. $10x^6 + 11x^5 - 29x^4 - 2x^3 - x^2 - 2x$

4. $6x^6 - 11x^5 + 25x^4 - 5x^3 + 19x^2 - 14x - 20$ 5. $2x^6 - 3x^5 + 6x^4 - 19x^3 + 15x^2 - 30x + 45$

6. $x - 2$ 7. $2x^2 - 5x + 8$ 8. $3x^2 - 5$ 9. $2x^2 - xy + 3y^2$ 10. $5x^2 - 2xy - y^2$

11. $2x^2 + 3x + 27$ remainder 126   12. $5x^2 - 8x + 5$ remainder -9 13 (a) $x = \dfrac{15}{2}$ and 2

(b) $x = -\dfrac{3}{2}$ and $\dfrac{1}{2}$ 14. $(x+1)(x-2)(x-3)$. Hence, $x = -1, 2$ or 3   15. $x = 1, -2, 3$ or $-\dfrac{5}{2}$ 16. 11

17. $7\dfrac{1}{4}$ 18. $2x+3$ and $2x+1$ 19. $x = 0, -\dfrac{1}{2}$ or $\dfrac{1}{2}$ 20. $3x^2 - 5$ 21. $a = 2, b = 8$

22 (a) $m = -9, n = 10$ (b) $x^2 - x - 10$ (c) $x = 1, 3.7$ or -2.7 23. $m = \dfrac{1}{2}, n = 39$

Exercise 10

1. $\dfrac{2}{x+3} + \dfrac{3}{x-5}$ 2. $\dfrac{17}{9(x-3)} - \dfrac{16}{9(2x+3)}$ 3. $\dfrac{1}{x-4} + \dfrac{1}{x+4}$ 4. $\dfrac{1}{x} + \dfrac{1}{2(x-5)}$ 5. $\dfrac{1}{x-2} + \dfrac{9}{(x-2)^2}$

6. $\dfrac{2}{x^2} - \dfrac{5}{2x+3}$ 7. $\dfrac{2}{x+1} + \dfrac{3}{(x-2)^2}$ 8. $\dfrac{2}{3x} + \dfrac{13x-3}{3(x^2+3)}$ 9. $\dfrac{2}{x^2+2} + \dfrac{3}{x+1}$

10. $\dfrac{3}{x-2} - \dfrac{5}{x+3} + \dfrac{2}{x-1}$ or $\dfrac{3}{x-2} + \dfrac{11-3x}{x^2+2x-3}$ 11. $\dfrac{2}{x+3} + \dfrac{3x-1}{x^2-3x-3}$

12. $\dfrac{1}{x-1} + \dfrac{3}{(x-1)^2} - \dfrac{5}{(x-1)^3}$ 13. $\dfrac{1}{x} + \dfrac{5}{x^2} + \dfrac{x-2}{x^2-1}$ 14. $1 - \dfrac{1}{4(x-3)} - \dfrac{3}{4(x+1)}$

15. $x + 5 + \dfrac{17}{5(x+4)} - \dfrac{7}{5(x-1)}$ 16. 1   17. $P = -\dfrac{11}{6}, Q = -\dfrac{13}{12}$ 18. $1 - x + \dfrac{4}{x-1} + \dfrac{3x-1}{x^2+1}$

Exercise 11

1. $x = 26$ 2. $x = 7$ 3. $x = 4\dfrac{1}{2}$ or $\dfrac{1}{2}$ 4. $x = 10$ 5. $x = 1$ 6. $x = 5$ 7. $x = 5$ 8. $x = 6$ or -6

9. $x = 4$ or -2 10. $x = 1$ or 3

Exercise 12

1. 9   2. $-\dfrac{8}{5}$ 3. 9   4. 14   5. -14 6. $\dfrac{2}{5}$ 7. $\dfrac{1}{3}$ 8. -1 9. -15

10. 10   11. $\dfrac{1}{4}$ 12. 27   13. 12   14. $\dfrac{1}{10}$ 15. $\dfrac{3}{7}$ 16. Continuous

17. Continuous   18. Discontinuous   19. Not continuous (Discontinuous

20. Not continuous   21. Continuous   22. Not continuous   23. Not continuous

24(a) 8   (b) 5   26. $\dfrac{7}{2}$ 27. $\dfrac{1}{3}$ 28. 1   29. $-\dfrac{1}{2}$ 30. $2\dfrac{1}{4}$

Exercise 13

1. 2   2. $2x$ 3. $-3x^{-4}$ or $-\dfrac{3}{x^4}$ 4(a) $10x + 5h$ (b) $10x$

5(a) $27x^2 + 27xh + 9h^2$ (b) $27x^2$ 6(a) $\dfrac{2x^3 + 3x^2\Delta x + x(\Delta x)^2 + 2}{x(x + \Delta x)}$ (b) $2x + \dfrac{2}{x^2}$

7. $6x - 10$ 8. $1 - \dfrac{3}{x^2}$ 9. $5 - 6x$ 10. $2 + \dfrac{1}{5} = 2\dfrac{1}{5}$

Exercise 14

1(a) $40x^4$ (b) $2x^4$ (c) $\dfrac{1}{3x^{\frac{2}{3}}}$ or $\dfrac{1}{3\sqrt[3]{x^2}}$ (d) $\dfrac{1}{x^{\frac{6}{7}}}$ or $\dfrac{1}{\sqrt[7]{x^6}}$ (e) $-\dfrac{5}{8x^{\frac{13}{8}}}$ or $-\dfrac{5}{8\sqrt[8]{x^{13}}}$

(f) $-\dfrac{5}{x^{\frac{7}{2}}}$ or $-\dfrac{5}{\sqrt{x^7}}$ 2. (a) $10x^4 - 12x^3 - 12x^2 + 10x - 6$ (b) $7x^6 + 8x^3 + \dfrac{3}{x^2}$

(c) $18x^5 - 4x^3 - 5 - \dfrac{2}{x^2} + \dfrac{3}{x^4}$ (d) $\dfrac{5}{4\sqrt[4]{x^3}} - \dfrac{5}{3\sqrt[3]{2x^4}}$ 3. $8(2x-5)^3$ 4. $\dfrac{-6}{(x^3-7)^3}$

5. $\dfrac{6x^2+7}{2(2x^3+7x)^{\frac{1}{2}}}$ 6. $5(21x^2-2x)(7x^3-x^2+3)^4$ 7. $\left(9+\dfrac{2}{x^2}\right)\left(3x-\dfrac{2}{3x}\right)^2$

8. $1 + \dfrac{9(6x-1)}{(3x^2-x-10)^2}$ 9. $\dfrac{-4x^3}{\sqrt{1-2x^4}}$ 10. $\dfrac{-5x^2}{\sqrt[3]{(5x^3-1)^4}}$

Exercise 15

1. $12x - 1$ 2. $12x^3 + 50x$ 3. $\dfrac{15x+30}{2\sqrt{3+x}}$ 4. $(7x^2 + 30 - 7)(x^2 - 7)^2$

5. $48x^3 + 24x^2 - 6x - 20$ 6. $\dfrac{\sqrt{2}\,[(3x^2-1)^3 + 36x^2(9x^4-6x^2+1)]}{2\sqrt{x}}$

7. $\dfrac{15-7x}{4(x+3)^{\frac{1}{4}}}$ or $\dfrac{15-7x}{4\sqrt[4]{x+3}}$ 8. $12x^3 + 9x^2 - 46x - 5$ 9. $15x^4 - 12x^3 - 3x^2 + 6x - 2$

10. $\dfrac{12x+1}{2}$ 11. $\dfrac{9x^{\frac{7}{2}}}{2}$ or $\dfrac{9\sqrt{x^7}}{2}$ 12. $\dfrac{4x^3(7x-33)}{3\sqrt[3]{2x-11}}$

13. $-30x^5 - 175x^4 + 20x^3 + 102x^2 - 14x + 1$ 14. $75x^4 + 28x^3 + 3x^2 + 28x - 7$

15. $-\dfrac{2}{x^2} + \dfrac{12}{x^5} - \dfrac{25}{x^6}$

Exercise 16

1. $\dfrac{x^2-2x+4}{(x-1)^2}$ 2. $\dfrac{8x^2+24x-3}{(2x+3)^2}$ 3. $\dfrac{42x}{(3x^2+1)^2}$ 4. $\dfrac{1}{(1-x)^{\frac{3}{2}}\sqrt{x+1}}$

5. $\dfrac{4(x^3-2)(x^3+1)}{x^3}$ 6. $\dfrac{-6(x^2+1)}{x^4\sqrt{3x^2+2}}$ 7. $\dfrac{-8x^2+12x+24}{3(2x+1)^3(x^2-x-4)^{\frac{2}{3}}}$ 8. $\dfrac{10x^3-2x^2-x+4}{2(2-x)^{\frac{3}{2}}}$

9. $\dfrac{-6x}{(1-3x^2)^2}$ 10. $\dfrac{4}{(x-2)^2}$ or $\dfrac{4}{(2-x)^2}$

Exercise 17

1. $\dfrac{3t}{2}$ 2. $\dfrac{2(4-t^3)^2}{3t}$ 3. $\dfrac{2r}{3(l+2r)}$ 4. $\dfrac{2t-1}{10t-1}$ 5. $\dfrac{-2m(v-u)}{(u+v)t^2}$ 6. $\dfrac{5(t^4+1)^2}{3(t^2-1)^2}$

7. $\dfrac{1-12t^2}{2t}$ 8. $\dfrac{15t^2}{2}$ 9. $\dfrac{8}{r}$ 10. $\dfrac{2s^3}{3}$

Exercise 18

1. $\dfrac{-15x^2}{3-2y}$ or $\dfrac{15x^2}{2y-3}$ 2. $\dfrac{-4x}{9y^2}$ 3. $\dfrac{3y^2+12x^2-y}{x-6xy}$ 4. $\dfrac{3x^2-y^2-4x}{2xy}$ 5. $\dfrac{x^2}{y}$

6. $\dfrac{4xy}{2x^2+5}$ 7. $\dfrac{-2xy}{x^2+6y^2-1}$ 8. $\dfrac{4y}{10y-4x-1}$ 9. $\dfrac{2xy^2-1}{2y(x^2-1)}$ 10. $\dfrac{-5x^4y^2}{3}$

Exercise 19

1. $2\sec^2 2x$ 2. $-\dfrac{1}{5}\sin\dfrac{1}{5}x$ 3. $-50\sin 5x$ 4. $3\sin^2 x\cos x$ 5. $2\sec^2 x\tan x$

6. $60x^4\sin^3 3x^5\cos 3x^5$ 7. $-12x\cot 6x^2\csc 6x^2$ 8. $10x^4\sec 2x^5\tan 2x^5$ 9. $6x\sec^2 3x^2$

10. $-x(3x\sin 3x - 2\cos 3x)$ 11. $\dfrac{6x\cos 2x - 9\sin 2x}{x^4}$ 12. $48x^3\sec x^4\tan x^4$

13. $\dfrac{-5\tan x\cos 5x - \sec^2 x(2-\sin 5x)}{\tan^2 x}$ 14. $\dfrac{\sin^2 x(3x\cos x - \sin x)}{2x^2}$

15. $\dfrac{-(2x+6)\csc^2 2x - \cot 2x)}{(x+3)^2}$ 16. $\dfrac{-\sin\sqrt[3]{3}}{5x^2}$ 17. $\dfrac{2[x-(x^2-3)\tan 2x]}{\sec 2x}$

18. $6x\cos 3x^2 - 18x^3\sin 3x^2$ 19. $-\cos x(2\cot x\sin x + \cos x\csc^2 x)$

20. $\dfrac{2(\sin^2 x + \sin^2 x)}{(\sin x+\cos x)^2}$ or $\dfrac{2}{(\sin x+\cos x)^2}$ 21. $\dfrac{4\sec 4x\sin(4x-2) + \tan 4x\cos(4x-2)}{\cos^2(4x-2)}$

22. $-3\sin 3x - x^2\sec x\tan x - 2x\sec x$ 23. $27x^2\cos x^3\sin^8 x^3$ 24. $\dfrac{3\cos 6\sqrt{x}}{\sqrt{x}}$

25. $3\cos x^3 - \dfrac{2\sin x^3}{x^3}$ 26. $-30x^4\cos^2 2x^5\sin 2x^2$ 27. $-10\sin 2x\sin 10x - 2\cos 2x\cos 10x$

28. $6x^2\cos x^4 - \dfrac{3\sin x^4}{2x^2}$ 29. $\dfrac{(10x-5)\sec^2 5x - 2\tan 5x}{(2x-1)^2}$ 30. $\dfrac{\sec^2 x\tan x}{x} - \dfrac{\tan^2 x}{2x^2}$

Exercise 20

1. $\dfrac{1}{3\sqrt[3]{7y^2}}$ 2. $\dfrac{3y^2}{2\sqrt{y^3-1}}$ 3. $\dfrac{1}{5\sqrt[5]{2(y+3)^4}}$ 4. $\dfrac{1}{\sqrt[3]{2(3y+27)^2}}$ 5. $\dfrac{-1}{(y-2)^2}$

6. $\dfrac{-3}{2y^2\sqrt{\dfrac{3}{y}+5}}$ 7. $2y^3$ 8. $5y^4$ 9. $\dfrac{-5}{(y-4)^2}$ 10. $\dfrac{-1}{3y^2\sqrt[3]{(\frac{1}{y}+8)^2}}$

Exercise 21

1. $\dfrac{2}{\sqrt{1-4x^2}}$ 2. $\dfrac{-1}{\sqrt{1-x^2}}$ 3. $\dfrac{-6x}{9x^4+1}$ 4. $\dfrac{-4x^3}{x^8+1}$ 5. $3\sec^{-1}x + \dfrac{3}{\sqrt{x^2-1}}$

6. $2x + \dfrac{3}{x^2+1}$ 7. $\dfrac{1}{(x+5)^2+1}$ or $\dfrac{1}{x^2+10x+26}$ 8. $\dfrac{1}{x^2\sqrt{1-\dfrac{1}{x^2}}}$ 9. $\dfrac{-1}{5\sqrt{1-y^2}}$

10. $\dfrac{1}{2y\sqrt{y-1}}$ 11. $5\tan^{-1}3x + \dfrac{15x}{9x^2+1}$ 12. $2x + \dfrac{20x^4}{\sqrt{1-x^{10}}}$ 13. $\dfrac{-3}{x\sqrt{x^6-1}}$

14. $\dfrac{1}{2(y^2+1)}$ 15. $\dfrac{1}{15y^{\frac{2}{3}}\sqrt{1-y^{\frac{2}{3}}}}$

307

Exercise 22

1. $3\cosh 3x - 2\sinh x$ 2. $-5x\operatorname{sech}x(x\tanh x - 2)$ 3. $\dfrac{2(x\cosh 2x - \sinh 2x)}{3x^3}$

4. $6\cosh 3x\sinh 3x$ 5. $60x^4\cosh^2 4x^5\sinh 4x^5$ 6. $-10x^4\operatorname{cosech}2x^5$

7. $-12x\operatorname{sech}^3 2x^2\tanh 2x^2$ 8. $x(5x\sinh 5x + 2\cosh 5x)$

9. $\dfrac{3\operatorname{cosech}^2 3x\tanh 5x + 5\coth 3x\operatorname{sech}^2 5x}{\coth^2 3x}$ 10. $\dfrac{2\cosh^2 x(3x\sinh x - \cosh x)}{3x^2}$

Exercise 23

1. $\dfrac{1}{x}\log_a e$ 2. $\dfrac{3x^2}{x^3+5}\log_a e$ 3. $\dfrac{36x^2}{2x^3-5}\log_a e$ 4. $\dfrac{2}{3x}\log_a e$ 5. $\dfrac{4x}{(x^2+1)(1-x^2)}\log_a e$

or $\dfrac{-4x}{(x^2+1)(x^2-1)}\log_a e$ or $\dfrac{-4x}{(x^4-1)}\log_a e$ 6. $\dfrac{15x^2}{5x^3-1}\log_5 e$ 7. $\dfrac{-2}{x}\log_2 e$ 8. $\dfrac{2x}{x^2+3}$

9. $\dfrac{3\ln^2 x}{x}$ 10. $\dfrac{5x^4}{2x^5-1}$ 11. $2x^3(4\ln x + 1)$ 12. $\dfrac{-28}{3-7x}$ or $\dfrac{28}{7x-3}$

13. $6x\ln(4x^3+1) + \dfrac{36x^4}{4x^3+1}$ 14. $\dfrac{2-4\ln x}{x^3}$ 15. $\dfrac{-30x}{1-5x^2}$ or $\dfrac{30x}{5x^2-1}$

Exercise 24

1. $10a^{2x}\log_e a$ 2. $a^{5x^2-x}\ln a(10x-1)$ 3. $a^{3x}x^3(3x\ln a + 4)$

4. $-2e^{-2x} - 3e^{-x}$ or $-e^{-2x}(2 + 3e^x)$ 5. $\dfrac{\sqrt{3}\,e^{\sqrt{3x}}}{\sqrt{x}}$ 6. $6x^2 e^{x^2}(2x^2+3)$

7. $\dfrac{5\ln a\,(a^{10x}-1)}{a^{5x}}$ 8. $\dfrac{6e^{4x}(2x\ln 2x^2 + 1)}{x}$ 9. $\dfrac{3x\sqrt{e^{5x}}(5x+4)}{2a}$ 10. $x^3(\log_{10}e + 4\log_{10}7x)$

11. $3\ln 6(6^{3x})$ 12. $2x - 3(e^{x^2-3x})$ 13. $\dfrac{-(5x+3)e^{\frac{3}{x}}}{5x^7}$ 14. $6e^{-3x}$

15. $3(2x+1)(x^2+x)^2 e^{(x^2+x)^3}$ 16. $2^{x^2}2x\ln 2$ 17. $2a^{2x}\ln a$ 18. $\dfrac{e^x(x\ln 10x^3 + 3)}{x}$

19. $\dfrac{\sqrt{5}(4x+1)e^{2x}}{2\sqrt{x}}$ 20. $2x^4(4\log_{10}e + 5\log_{10}5x^4)$

Exercise 25

1. $x^{2x}(2 + 2\ln x)$ 2. $\dfrac{3x^{\ln 2x}(\ln 2x + \ln x)}{x}$ 3. $\ln(1-4x^2) - \dfrac{8x^2}{1-4x^2}$

4. $(6x^x - 5)^{3x}\left[3\ln(6x^2-5) + \dfrac{36x^2}{6x^2-5}\right]$ 5. $\dfrac{(2x^3+1)(x-2)^3}{3x^2(x^3-1)^2}\left(\dfrac{6x^2}{2x^3+1} + \dfrac{3}{x-2} - \dfrac{2}{x} - \dfrac{6x^2}{x^3-1}\right)$

6. $\dfrac{(2x-1)(x^2-2)}{(1-x)(x-3)^2}\left(\dfrac{2}{2x-1} + \dfrac{2x}{x^2-2} + \dfrac{1}{1-x} - \dfrac{2}{x-3}\right)$

7. $15x^3 x^{3x^4}(4\ln x + 1)$ or $15x^{3x^4+3}(4\ln x + 1)$ 8. $3e^{3x}(e^{e^{3x}})$ or $3e^{e^{3x}+3x}$

9. $2e^x e^x \ln 2$ 10. $\dfrac{x^{\ln(x^2-4x)}[(x-4)\ln(x^2-4x) + (2x-4)\ln x]}{x(x-4)}$ 11. $\ln(2x^2+10) + \dfrac{4x^2}{2x^2+10}$

308

12. $(3x^2 - 8)^{2x}\left[2\ln(3x^3 - 8) + \dfrac{18x^3}{3x^3 - 8}\right]$ 13. $\dfrac{(x+3)^2(x-3)^2}{x^3-1}\left(\dfrac{2}{x+3} + \dfrac{2}{x-3} - \dfrac{3x^2}{x^3-1}\right)$

14. $15x^{(\ln x)^2}(\ln x)^2$ 15. $30x^{3x^2}x(2\ln x + 1)$ or $30x^{3x^2+1}(2\ln x + 1)$

Exercise 26

1. $\dfrac{3}{5x^2}$ 2. $\dfrac{5xe^{5x}}{2}$ 3. $\dfrac{-2x\sin x^2}{\cos x}$ 4. $\dfrac{x(6x-5)}{x-1}$ 5. $\dfrac{14x^2}{3}$ 6. $\dfrac{2a^x\ln a}{e^x}$

7. $\dfrac{-2\tan x\, \sec^2 x}{5\sin 5x}$ 8. $\dfrac{1}{(x-1)(e^{x-1})}$ 9. $\dfrac{1}{xe^{5x}}$ 10. $\dfrac{e^x + e^{-x}}{(e^x - e^{-x})(1 + \ln x)}$

Exercise 27

1. $\dfrac{dy}{dx} = 2x(2x^2 - 3x - 7)$, $\dfrac{d^2y}{dx^2} = 2(6x^2 - 6x - 7)$, $\dfrac{d^3y}{dx^3} = 2(12x - 6)$ 2. $-\dfrac{3}{x^2}$

3. $20x^2e^{5x^4}(20x^4 + 3)$ 4. $-50x^8\sin^2 x^5 + 40x^3\cos x^5\sin x^5 + 50x^8\cos x^5$ 5. $\dfrac{128}{(4x-10)^2}$

6. $\dfrac{-(x^2-2)\sin x - 2x\cos x}{x^3}$ 7. $\dfrac{4\ln x - 6}{x^3}$ 8. $-2\csc^2 x(\csc^2 x + 2\cot^2 x)$

9. $2\cos^2 x - 3\sin^2 x - 5\cos 2x$ 10. 0

11. $-8x^2\sin x^2 + 4\cos x^2 + 2\sin^2 x - 2\cos^2 x - 4x\cos x^2 + 2\cos x\sin x$

12. $\dfrac{2[(x^2\sec^2 x + 1)\tan x - x\sec^2 x]}{3x^3}$ 13. $e^{-x}(4\sin 2x - 3\cos 2x)$ 14. 0

15. $\dfrac{-(9x^2-2)\sin 3x - 6x\cos 3x}{3x^3}$

Exercise 28

1. $2x(3x^4 - 5)^3 + 36x(3x^4 - 5)^3$ or $2x(3x^4 - 5)^2(21x^4 - 5)$ 2. 209

3(a) $\dfrac{\sqrt{3x^2-1}\,(3x^2+2)}{x^3}$ (b) $5\sqrt{2}$ 4(a) $\dfrac{12x^2 - 3y^3}{9xy^2 - 2y - 5}$ (b) $-\dfrac{9}{16}$ 5. $-\dfrac{2}{y}$

6. $3x(x^2 + 5)^2(3x^3 + 5x - 2)$ 7. $e^{\sin x + \tan x}(\sec^2 x + \cos x)$ 8. $\dfrac{2x(\sin x - x\cos x)}{\sin^3 x}$

9. $\dfrac{2\cos x\sin x - 3\sin 3x}{\sin^2 x + \cos 3x}$ 10. $-5a^{\cos 5x}\sin 5x\ln a$ 11. $5x^2 e^{-3x} - 10xe^{-3x} + 2e^{-3x}$

12. $-(2x^3 + 2)\cos\left(\dfrac{1-2x^3}{x^2}\right)$ 13. $\dfrac{4x^4 - 6x^3 + 12x^2 - 18x + 6}{x^3}$ 14. 0

15. $5\left(\dfrac{2}{x^2} - \dfrac{2}{x^3}\right)\left(\dfrac{1}{x^2} - \dfrac{2}{x}\right)^4$ or $\dfrac{10(x-1)(1-2x)^4}{x^{11}}$ 16(a) $\dfrac{3x^2e^{x^3-y^3} - y}{3y^2e^{x^3-y^3} + x}$ (b) 1

17. 0 18. $\dfrac{2\sin\frac{1}{x^2}}{x^2} - \dfrac{4\cos\frac{1}{x^2}}{x^4} + 2\cos\dfrac{1}{x^2}$ 19(a) $\dfrac{2e^x(e^x + 1)}{(e^x - 1)^3}$

(b) $\dfrac{4(e^{2x} - e^{-x}) + (1 + e^{-x})(e^x - 1)^3}{(e^x - 1)^3(1 - e^{-x})}$ 20. $2\tan x$ 21. 0

22. $3\left(\dfrac{\sqrt{5}}{4\sqrt{x}} + 3\right)\left(3x + \dfrac{\sqrt{5x}}{2}\right)^2$

23. $\dfrac{-10x - 18}{x^4}$ or $\dfrac{-2(5x + 9)}{x^4}$

24. $9x^2(x^3 + 1)^2 + \dfrac{5 + 4x - 10x^2}{(x^2 + 1)^2}$

25. 0

Please if you found this book well simplified enough for easier understanding, kindly give it a five star rating on amazon so as to encourage people to buy this book, thereby helping them improve on their skills on algebra and differential calculus. Thank you.

If you want to see other books written by the author, just simply search for the author's name, Kingsley Augustine on amazon.com

If you have any enquiry, suggestion or information concerning this book, please contact the author through the email below.

KINGSLEY AUGUSTINE

kingzohb2@yahoo.com

www.ingramcontent.com/pod-product-compliance
Lightning Source LLC
Chambersburg PA
CBHW082208290526
45794CB00009B/3469